Physics & Philosophy

THE REVOLUTION IN MODERN SCIENCE

This Torchbook paperback edition reprints Volume XIX of the WORLD PERSPECTIVES SERIES, which is planned and edited by RUTH NANDA ANSHEN. Dr. Anshen's Epilogue to this reprint appears on page 209 ff.

Physics & Philosophy

THE REVOLUTION IN MODERN SCIENCE

WERNER HEISENBERG

HARPER TORCHBOOKS ▼ THE SCIENCE LIBRARY

HARPER & ROW, PUBLISHERS

NEW YORK

PHYSICS AND PHILOSOPHY

Printed in the United States of America

First HARPER TORCHBOOK *edition published 1962*

Library of Congress catalog card number: 58-6150

88 89 90 26 25 24 23 22 21

Contents

Introduction

by F. S. C. Northrop

Sterling Professor of Philosophy and Law,
The Law School, Yale University

There is a general awareness that contemporary physics has brought about an important revision in man's conception of the universe and his relation to it. The suggestion has been made that this revision pierces to the basis of man's fate and freedom, affecting even his conception of his capacity to control his own destiny. In no portion of physics does this suggestion show itself more pointedly than in the principle of indeterminacy of quantum mechanics. The author of this book is the discoverer of this principle. In fact, it usually bears his name. Hence, no one is more competent to pass judgment on what it means than he.

In his previous book, *The Physical Principles of the Quantum Theory,** Heisenberg gave an exposition of the theoretical interpretation, experimental meaning and mathematical apparatus of quantum mechanics for professional physicists. Here he pursues this and other physical theories with respect to their philosophical implications and some of their likely social consequences for the layman. More specifically, he attempts here to raise and suggest answers to three questions: (1) What do the experimentally verified theories of contemporary physics affirm? (2) How do they permit or require man to think of himself in relation to his universe? (3) How is this new way of

*** University of Chicago Press, Chicago, 1930.**

thinking, which is the creation of the modern West, going to affect other parts of the world?

The third of these questions is dealt with briefly by Heisenberg at the beginning and end of this inquiry. The brevity of his remarks should not lead the reader to pass lightly over their import. As he notes, whether we like it or not, modern ways are going to alter and in part destroy traditional customs and values. It is frequently assumed by native leaders of non-Western societies, and also often by their Western advisers, that the problem of introducing modern scientific instruments and ways into Asia, the Middle East and Africa is merely that of giving the native people their political independence and then providing them with the funds and the practical instruments. This facile assumption overlooks several things. First, the instruments of modern science derive from its theory and require a comprehension of that theory for their correct manufacture or effective use. Second, this theory in turn rests on philosophical, as well as physical, assumptions. When comprehended, these philosophical assumptions generate a personal and social mentality and behavior quite different from, and at points incompatible with, the family, caste and tribally centered mentality and values of the native Asian, Middle Eastern or African people. In short, one cannot bring in the instruments of modern physics without sooner or later introducing its philosophical mentality, and this mentality, as it captures the scientifically trained youth, upsets the old familial and tribal moral loyalties. If unnecessary emotional conflict and social demoralization are not to result, it is important that the youth understand what is happening to them. This means that they must see their experience as the coming together of two different philosophical mentalities, that of their traditional culture and that of the new physics. Hence, the importance for everyone of understanding the philosophy of the new physics.

But it may be asked, Isn't physics quite independent of philosophy? Hasn't modern physics become effective only by dropping philosophy? Clearly, Heisenberg answers both of these questions in the negative. Why is this the case?

Newton left the impression that there were no assumptions in his physics which were not necessitated by the experimental data. This occurred when he suggested that he made no hypotheses and that he had deduced his basic concepts and laws from the experimental findings. Were this conception of the relation between the physicist's experimental observations and his theory correct, Newton's theory would never have required modification, nor could it ever have implied consequences which experiment does not confirm. Being implied by the facts, it would be as indubitable and final as they are.

In 1885, however, an experiment performed by Michelson and Morley revealed a fact which should not exist were the theoretical assumptions of Newton the whole truth. This made it evident that the relation between the physicist's experimental facts and his theoretical assumptions is quite other than what Newton had led many modern physicists to suppose. When, some ten years later, experiments on radiation from black bodies enforced an additional reconstruction in Newton's way of thinking about his subject matter, this conclusion became inescapable. Expressed positively, this means that the theory of physics is neither a mere description of experimental facts nor something deducible from such a description; instead, as Einstein has emphasized, the physical scientist only arrives at his theory by speculative means. The deduction in his method runs not from facts to the assumptions of the theory but from the assumed theory to the facts and the experimental data. Consequently, theories have to be proposed speculatively and pursued deductively with respect to their many consequences so that they can be put to indirect experimental tests. In short, any theory of physics makes more physical and philosophical assumptions than the facts alone give or imply. For this reason, any theory is subject to further modification and reconstruction with the advent of new evidence that is incompatible, after the manner of the results of the Michelson-Morley experiment, with its basic assumptions.

These assumptions, moreover, are philosophical in character. They may be ontological, i.e., referring to the subject matter of scientific knowledge which is independent of its relation to the perceiver; or they may be epistemological, i.e., referring to the relation of the scientist as experimenter and knower to the subject matter which he knows. Einstein's special and general theories of relativity modify the philosophy of modern physics in the first of these two respects by radically altering the philosophical theory of space and time and their relation to matter. Quantum mechanics, especially its Heisenberg principle of indeterminacy, has been notable for the change it has brought in the physicist's epistemological theory of the relation of the experimenter to the object of his scientific knowledge. Perhaps the most novel and important thesis of this book is its author's contention that quantum mechanics has brought the concept of potentiality back into physical science. This makes quantum theory as important for ontology as for epistemology. At this point, Heisenberg's philosophy of physics has an element in common with that of Whitehead.

It is because of this introduction of potentiality into the subject matter of physics, as distinct from the epistemological predicament of physicists, that Einstein objected to quantum mechanics. He expressed this objection by saying: "God does not play dice." The point of this statement is that the game of dice rests on the laws of chance, and Einstein believed that the latter concept finds its scientific meaning solely in the epistemological limitations of the finite knowing mind in its relation to the omnicomplete object of scientific knowledge and, hence, is misapplied when referred ontologically to that object itself. The object being per se all complete and in this sense omniscient, after the manner of God, the concept of chance or of probability is inappropriate for any scientific description of it.

This book is important because it contains Heisenberg's answer to this criticism of his principle of indeterminacy and of quantum theory by Einstein and by others. In understanding this answer two things must be kept in mind: (1) The aforementioned relation between the data of experimental physics and the concepts of its theory. (2) The difference between the role of the concept of probability in (a) Newton's mechanics and Einstein's theory of relativity and in (b) quantum mechanics. Upon (1), Einstein and Heisenberg, and relativistic mechanics and quantum mechanics, are in agreement. It is only with respect to (2) that they differ. Yet the reason for Heisenberg's and the quantum physicist's difference from Einstein on (2) depends in considerable part on (1) which Einstein admits.

(1) affirms that the experimental data of physics do not imply its theoretical concepts. From this it follows that the object of scientific knowledge is never known directly by observation or experimentation, but is only known by speculatively proposed theoretic construction or axiomatic postulation, tested only indirectly and experimentally via its deduced consequences. To

find the object of scientific knowledge we must go, therefore, to its theoretical assumptions.

When we do this for (a) Newton's or Einstein's mechanics and for (b) quantum mechanics, we discover that the concept of probability or chance enters into the definition of the state of a physical system, and, in this sense, into its subject matter, in quantum mechanics, but does not do so in Newton's mechanics or Einstein's theory of relativity. This undoubtedly is what Heisenberg means when he writes in this book that quantum theory has brought the concept of potentiality back into physical science. It is also, without question, what Einstein has in mind when he objects to quantum theory.

Put more concretely, this difference between quantum mechanics and previous physical theories may be expressed as follows: In Newton's and Einstein's theory, the state of any isolated mechanical system at a given moment of time is given precisely when only numbers specifying the position and momentum of each mass in the system are empirically determined at that moment of time; no numbers referring to a probability are present. In quantum mechanics the interpretation of an observation of a system is a rather complicated procedure. The observation may consist in a single reading, the accuracy of which has to be discussed, or it may comprise a complicated set of data, such as the photograph of the water droplets in a cloud chamber; in any case, the result can be stated only in terms of a probability distribution concerning, for instance, the position or momentum of the particles of the system. The theory then predicts the probability distribution for a future time. The theory is not experimentally verified when that future state arrives if merely the momentum or position numbers in a particular observation lie within the predicted range. The same experiment with the same initial conditions must be repeated many times, and the values of position or momentum, which may be different in each observation, must similarly be found to be distributed according to the predicted probability distribution. In short, the crucial difference between quantum mechanics and Einstein's or Newton's mechanics centers in the definition of a mechanical system at any moment of time, and this difference is that quantum mechanics introduces the concept of probability into its definition of state and the mechanics of Newton and Einstein does not.

This does not mean that probability had no place in Newton's or Einstein's mechanics. Its place was, however, solely in the theory of errors by means of which the accuracy of the Yes or No verification or nonconfirmation of the prediction of the theory was determined. Hence, the concept of probability and chance was restricted to the epistemological relation of the scientist in the verification of what he knows; it did not enter into the theoretical statement of what he knows. Thus, Einstein's dictum that "God does not play dice" was satisfied in his two theories of relativity and in Newton's mechanics.

Is there any way of deciding between Einstein's contention and that of Heisenberg and other quantum theorists? Many answers have been given to this question. Some physicists and philosophers, emphasizing operational definitions, have argued that, since all physical theories, even classical ones, entail human error and uncertainties, there is nothing to be decided between Einstein and the quantum theorists. This, however, is (a) to overlook the presence of axiomatically constructed, constitutive theoretic definitions as well as theory-of-errors, operational definitions in scientific method and (b) to suppose that the concept of probability and the even more complex uncertainty relation enter into quantum mechanics only in the operational-definitional sense. Heisenberg shows that the latter supposition is false.

Other scientists and philosophers, going to the opposite extreme, have argued that, merely because there is uncertainty in predicting certain phenomena, this constitutes no argument whatever for the thesis that these phenomena are not completely determined. This argument combines the statical problem of defining the state of a mechanical system at a given time with the dynamical or causal problem of predicting changes in the state of the system through time. But the concept of probability in quantum theory enters only into its statics, i.e., its theoretical definition of state. The reader will find it wise, therefore, to keep distinct these two components, i.e., the statical theoretical definition-of-state component and the dynamic, or causal, theoretical change-of-state-through-time component. With respect to the former, the concept of probability and the attendant uncertainty enter theoretically and in principle; they do not refer merely to the operational and epistemological uncertainties and errors, arising from the finiteness of, and inaccuracies in, human behavior, that are common to any scientific theory and any experimentation whatsoever.

But, why, it may be asked, should the concept of probability be introduced into the theoretic definition of the state of a mechanical system at any statical moment t^1 *in principle?* In making such a theoretical construct by axiomatic postulation, do not Heisenberg and quantum theoreticians generally beg the question at issue between themselves and Einstein? This book makes it clear that the answer to these questions is as follows: The reason for the procedure of quantum mechanics is thesis (1) above, which Einstein himself also accepts.

Thesis (1) is that we know the object of scientific knowledge only by the speculative means of axiomatic theoretic construction or postulation; Newton's suggestion that the physicist can deduce our theoretical concepts from the experimental data being false. It follows that there is no *a priori* or empirical meaning for affirming that the object of scientific knowledge, or, more specifically, the state of a mechanical system at a given time t^1, must be defined in a particular way. The sole criterion is, which set of theoretic assumptions concerning the subject matter of mechanics when pursued to their deduced experimental consequences is confirmed by the experimental data?

Now, it happens that when we theoretically and in principle define the state of a mechanical system for subatomic phenomena in terms solely of numbers referring to position and momentum, as Einstein would have us do, and deduce the consequences for radiation from black bodies, this theoretical assumption concerning the state of a mechanical system and the subject matter of atomic physics is shown to be false by experimental evidence. The experimental facts simply are not what the theory calls for. When, however, the traditional theory is modified with the introduction of Planck's constant and the addition in principle of the second set of numbers referring to the probability that the attached position-momentum numbers will be found, from which the uncertainty principle follows, the experimental data confirm the new theoretical concepts and principles. In short, the situation in quantum mechanics with respect to experiments on black-body radiation is identical with that faced by Einstein with respect to the Michelson-Morley experiment. In both cases, only by introducing the new theoretical assumption in principle is physical theory brought into accord with the experimental facts. Thus, to assert that, notwithstanding quantum mechanics, the positions and momenta of subatomic masses are "really" sharply located in space and time as designated by one pair of numbers only and, hence, completely deterministic causally, as Einstein and the aforementioned philosophers of science would have one do, is to affirm a theory concerning the subject matter of physical knowledge which experiments on black-body radiation have shown to be false in the sense that a deductive experimental consequence of this theory is not confirmed.

It does not follow, of course, that some new theory compatible with the foregoing experimental facts might not be discovered in which the concept of probability does not enter in principle into its definition of state. Professor Norbert Wiener, for example, believes that he has clues to the direction such a theory might take. It would, however, have to reject a definition of state in terms of the four space-time dimensions of Einstein's theory and would, therefore, be incompatible with Einstein's thesis on other grounds. Certainly, one cannot rule out such a possibility. Nevertheless, until such an alternative theory is presented, anyone, who does not claim to possess some *a priori* or private source of information concerning what the object of scientific knowledge must be, has no alternative but to accept the definition of state of quantum theory and to affirm with the author of this book that it restores the concept of potentiality to the object of modern scientific knowledge. Experiments on black-body radiation require one to conclude that God plays dice.

What of the status of causality and determinism in quantum mechanics? Probably the interest of the layman and the humanist in this book depends most on its answer to this question.

If this answer is to be understood, the reader must pay particular attention to Heisenberg's description of (a) the aforementioned definition of state by recourse to the concept of probability and (b) the Schrödinger time-equation. The reader must also make sure, and this is the most difficult task of all, that the meaning of the words "causality" and "determinism" in his mind when he asks the above question is identical with the meaning these words have in Heisenberg's mind when he specifies the answer. Otherwise, Heisenberg will be answering a different question from the one the reader is asking and a complete misunderstanding upon the reader's part will occur.

The situation is further complicated by the fact that modern physics permits the concept of causality to have two different, scientifically precise meanings, the one stronger than the other, and there is no agreement among physicists about which one of these two meanings the word "causality" is to be used to designate. Hence, some physicists and philosophers of science use the word to designate the stronger of the two meanings. There is evidence, at times at least, that this is Professor Heisenberg's usage in this book. Other physicists and philosophers, including the writer of this Introduction, use the word "causality" to designate the weaker of the two meanings and the word "determinism" to designate the stronger meaning. When the former usage is followed, the words "causality" and "determinism" become synonymous. When the second usage is followed, every deterministic system is a causal system, but not every causal system is deterministic.

Great confusion has entered into previous discussion of this topic because frequently neither the person who asks the question nor the physicist who has answered it has been careful to specify in either question or answer whether he is using the word "causality" in its weaker or in its stronger modern scientific meaning. If one asks "Does causality hold in quantum mechanics?" not specifying whether one is asking about causality in its stronger or in its weaker sense, one then gets apparently contradictory answers from equally competent physicists. One physicist, taking the word "causality" in its stronger sense, quite correctly answers "No." The other physicist, taking "causality" in its weaker sense, equally correctly answers "Yes." Naturally, the impression has arisen that quantum mechanics is not specific about what the answer is. Nevertheless, this impression is erroneous. The answer of quantum mechanics becomes unequivocal the moment one makes the question and the answer unambiguous by specifying which meaning of "causality" one is talking about.

It is important, therefore, to become clear about different possible meanings of the word "causality." Let us begin with the layman's common-sense usage of the word "cause" and then move to the more exact meanings in modern physics, considering the meaning in Aristotle's physics on the way.

One may say "The stone hit the window and *cause*d the glass to break." In this use of "causality" it is thought of as a relation between objects, i.e., between the stone and the windowpane. The scientist expresses the same thing in a different way. He describes the foregoing set of events in terms of the state of the stone and the windowpane at the earlier time t^1 when the stone and the windowpane were separated and the state of this same system of two objects at the later time t^2 when the stone and the windowpane collided. Consequently, whereas the layman tends to think of causality as a relation between objects, the scientist thinks of it as a relation between different states of the same object or the same system of objects at different times.

This is why, in order to determine what quantum mechanics says about causality, one must pay attention to two things: (1) The state-function which defines the state of any physical system at any specific time t. (2) The Schrödinger time-equation which relates the state of the physical system at the earlier time t^1 to its different state at any specifiable later time t^2. What Heisenberg says about (1) and (2) must, therefore, be read with meticulous care.

It will help to understand what quantum mechanics says about the relation between the states of a given physical object, or system of physical objects, at different times if we consider the possible properties that this relation might have. The weakest possible case would be that of mere temporal succession with no necessary connection whatever and with not even a probability, however small, that the specifiable initial state will be followed in time by a specifiable future state. Hume gives us reasons for believing that the relation between the sensed states of immediately sensed natural phenomena is of this character. Certainly, as he pointed out, one does not sense any relation of necessary connection. Nor does one directly sense probability. All that sensation gives us with respect to the successive states of any phenomenon is the mere relation of temporal succession.

This point is of great importance. It means that one can arrive at a causal theory in any science or in common-sense knowledge, or even at a probability theory, of the relation between the successive states of any object or system, only by speculative means and axiomatically constructed, deductively formulated scientific and philosophical theory which is tested not directly against the sensed and experimental data but only indirectly by way of its deductive consequences.

A second possibility with respect to the character of the relation between the states of any physical system at different times is that the relation is a necessary one, but that one can know what this necessary connection is only by knowing the future state. The latter knowledge of the future state may be obtained either by waiting until it arrives or by having seen the future or final state of similar systems in the past. When such is the case, causality is teleological. Changes of the system with time are determined by the final state or goal of the system. The physical system which is an acorn in the earlier state t^1 and an oak tree in the later state t^2 is an example. The connection between these two states seems to be a necessary one. Acorns never change into maple trees or into elephants. They change only into oaks. Yet, given the properties of this physical system in the acorn state of the earlier time t^1, no scientist has as yet been able to deduce the properties of the oak tree which the system will have at the later time t^2. Aristotelian physics affirmed that all causal relations are teleological.

Another possibility is that the relation between the states of any object, or any system of objects, at different times is a relation of necessary connection such that, given knowledge of the initial state of the system, assuming isolation, its future state can be deduced. Stated in more technical mathematical language, this means that there exists an indirectly verified, axiomatically constructed theory whose postulates (1) specify a state-function, the independent variables of which completely define the state of the system at any specific instant of time, and (2) provide a time-equation relating the numerical empirical values of the independent variables of this function at any earlier time t^1 to their numerical empirical values at any specific later time t^2 in such a way that by introducing the operationally determined t^1 set of numbers into the time-equation the future t^2 numbers can be deduced by merely solving the equation. When this is the case, the temporal relation between states is said to exemplify mechanical causation.

It is to be noted that this definition of mechanical causality leaves open the question of what independent variables are required to define the state of the system at any given time. Hence, at least two possibilities arise: (a) the concept of probability may be used to define the state of the system or (b) it may not be so used. When (b) is the case no independent variables referring to probabilities appear in the state-function and the stronger type of mechanical causality is present. When (a) is the case independent variables referring to probabilities, as well as to other properties such as position and momentum, appear in the state-function and only the weaker type of mechanical causation occurs. If the reader keeps these two meanings of mechanical causation in mind and makes sure which meaning Heisenberg is referring to in any particular sentence of this book, he should be able to get its answer to the question concerning the status of causality in modern physics.

What of determinism? Again, there is no agreed-upon convention among physicists and philosophers of science about how this word is to be used. It is in accord with the common-sense usage to identify it with the strongest possible causality. Let us, then, use the word "determinism" to denote only the stronger type of mechanical causation. Then I believe the careful reader of this book will get the following answer to his question: In Newtonian, Einsteinian and quantum mechanics, mechanical, rather than teleological, causality holds. This is why quantum physics is called quantum mechanics, rather than quantum teleologics. But, whereas causality in Newton's and Einstein's physics is of the stronger type and, hence, both mechanical and deterministic, in quantum mechanics it is of the weaker causal type and, hence, mechanical but not deterministic. From the latter fact it follows that if anywhere in this book Heisenberg uses the words "mechanical causality" in their stronger, deterministic meaning and the question be asked "Does mechanical causation in this stronger meaning hold in quantum mechanics?" then the answer has to be "No."

The situation is even more complicated, as the reader will find, than even these introductory distinctions between the different types of causation indicate. It is to be hoped, however, that this focusing of attention upon these different meanings will enable the reader to find his way through this exceptionally important book more easily than would otherwise be the case.

These distinctions should suffice, also, to enable one to grasp the tremendous philosophical significance of the introduction of the weaker type of mechanical causation into modern physics, which has occurred in quantum mechanics. Its significance consists in reconciling the concept of objective, and in this sense ontological, potentiality of Aristotelian physics with the concept of mechanical causation of modern physics.

It would be an error, therefore, if the reader, from Heisenberg's emphasis upon the presence in quantum mechanics of something analogous to Aristotle's concept of potentiality, concluded that contemporary physics has taken us back to Aristotle's physics and ontology. It would be an equal error conversely to conclude, because mechanical causation in its weaker meaning still holds in quantum mechanics, that all is the same now in modern physics with respect to its causality and ontology as was the case before quantum mechanics came into being. What has occurred is that in quantum theory contemporary man has moved on beyond the classical medieval and the modern world to a new physics and philosophy which combines consistently some of the basic causal and ontological assumptions of each. Here, let it be recalled, we use the word "ontological" to denote any experimentally verified concept of scientific theory which refers to the object of scientific knowledge rather than merely to the epistemological relation of the scientist as knower to the object which he knows. Such an experimentally verified philosophical synthesis of ontological potentiality with ontological mechanical causality, in the weaker meaning of the latter concept, occurred when physicists found it impossible to account theoretically for the Compton effect and the results of experiment on black-body radiation unless they extended the concept of probability from its Newtonian and Einsteinian merely epistemological, theory-of-errors role in specifying when their theory is or is not experimentally confirmed to the ontological role, specified in principle in the theory's postulates, of characterizing the object of scientific knowledge itself.

Need one wonder that Heisenberg went through the subjective emotional experiences described in this book before he became reconciled to the necessity, imposed by both experimental and mathematical considerations, of modifying the philosophical and scientific beliefs of both medieval and modern man in so deep-going a manner. Those interested in a firsthand description of the human spirit in one of its most creative moments will want to read this book because of this factor alone. The courage which it took to make this step away from the unqualified determinism of classical modern physics may be appreciated if one recalls that even such a daring, creative spirit as Einstein balked. He could not allow God to play dice; there could not be potentiality in the object of scientific knowledge, as the weaker form of mechanical causality in quantum mechanics allows.

Before one concludes, however, that God has become a complete gambler and that potentiality is in all objects, certain limitations which quantum mechanics places on the application of its weaker form of mechanical causation must be noted. To appreciate these qualifications the reader must note what this book says about (1) the Compton effect, (2) Planck's constant h, and (3) the uncertainty principle which is defined in terms of Planck's constant.

This constant h is a number referring to the quantum of action of any object or system of objects. This quantum, which extends atomicity from matter and electricity to light and even to energy itself, is very small. When the quantum numbers of the system being observed are small, as is the case with subatomic phenomena, then the uncertainty specified by the Heisenberg uncertainty principle of the positions and momenta of the masses of the system becomes significant. Then, also, the probability numbers associated with the position-momentum numbers in the state-function become significant. When, however, the quantum numbers of the system are large, then the quantitative amount of uncertainty specified by the Heisenberg principle becomes insignificant and the probability numbers in the state-function can be neglected. Such is the case with gross common-sense objects. At this point quantum mechanics with its basically weaker type of causality gives rise, as a special case of itself, to Newtonian and Einsteinian mechanics with their stronger type of causality and determinism. Consequently, for human beings considered *merely* as gross common-sense objects the stronger type of causality holds and, hence, determinism reigns also.

Nevertheless, subatomic phenomena are scientifically significant in man. To this extent, at least, the causality governing him is of the weaker type, and he embodies both mechanical fate and potentiality. There are scientific reasons for believing that this occurs even in heredity. Any reader who wants to pursue this topic beyond the pages of this book should turn to *What Is Life?** by Professor Erwin Schrödinger, the physicist after whom the time-equation in quantum mechanics is named. Undoubtedly, potentiality and the weaker form of causality hold also for countless other characteristics of human beings, particularly for those cortical neural phenomena in man that are the epistemic correlates of directly introspected human ideas and purposes.

If the latter possibility is the case, the solution of a baffling scientific, philosophical and even moral problem may be at hand. This problem is: How is the mechanical causation, even in its weaker form, of quantum mechanics to be reconciled with the teleological causation patently present in the moral, political and legal purposes of man and in the teleological causal determination of his bodily behavior, in part at least, by these purposes? In short, how is the philosophy of physics expounded in this book by Heisenberg to be reconciled with moral, political and legal science and philosophy?

It may help the reader to appreciate why this book must be mastered before these larger questions can be correctly understood or effectively answered if very brief reference is made here to some articles which relate its theory of physical causation to the wider relation between mechanism and teleology in the humanities and the social sciences. The relevant articles are (a) by Professors Rosenblueth, Wiener and Bigelow in the journal of *The Philosophy of Science* for January, 1943; (b) by Doctors McCulloch and Pitts in *The Bulletin of Mathematical Biophysics,* Volume 5, 1943, and Volume 9, 1947; and (c) Chapter XIX of *Ideological Differences and World Order,* edited by the writer of this Introduction and published by the Yale University

*** University Press, Cambridge; Macmillan Company, New York; 1946.**

Press in 1949. If read after this book, (a) will show how teleological causality arises as a special case of the merchanical causality described by Heisenberg here. Similarly, (b) will provide a physical theory of the neurological correlates of introspected ideas, expressed in terms of the teleologically mechanical causality of (a), thereby giving an explanation of how ideas can have a causally significant effect on the behavior of men. Likewise, (c) will show how the ideas and purposes of moral, political and legal man relate, by way of (b) and (a), to the theory of physical potentiality and mechanical causality so thoroughly described by Heisenberg in this book.

It remains to call attention to what Professor Heisenberg says about Bohr's principle of complementarity. This principle plays a great role in the interpretation of quantum theory by "the Copenhagen School" to which Bohr and Heisenberg belong. Some students of quantum mechanics, such as Margenau in his book *The Nature of Physical Reality,** are inclined to the conclusion that quantum mechanics requires merely its definition of state, its Schrödinger time-equation and those other of its mathematical postulates which suffice to ensure, as noted above, that Einsteinian and Newtonian mechanics come out of quantum mechanics as one of its special cases. According to the latter thesis, the principle of complementarity arises from the failure to keep the stronger and weaker form of mechanical causality continuously in mind, with the resultant attribution of the stronger form to those portions of quantum mechanics where only the weaker form is involved. When this happens, the principle of complementarity has to be introduced to avoid

* McGraw-Hill Book Co., Inc., New York, 1950, pp. 418–422. See also, Northrop, *The Logic of the Sciences and the Humanities,* Macmillan, New York, 1947, Chapter XI.

contradiction. If, however, one avoids the foregoing practice, the principle of complementarity becomes, if not unnecessary, at least of a form such that one avoids the danger, noted by Margenau * and appreciated by Bohr, of giving pseudo solutions to physical and philosophical problems by playing fast and loose with the law of contradiction, in the name of the principle of complementarity.

By its use the qualifications that had to be put on both the particle-picture common-sense language of atomic physics and its common-sense wave-picture language were brought together. But once having formulated the result with axiomatically constructed mathematical exactitude, any further use of it is merely a superficial convenience when, leaving aside the exact and essential mathematical assumptions of quantum mechanics, one indulges in the common-sense language and images of waves and particles.

It has been necessary to go into the different interpretations of the principle of complementarity in order to enable the reader to pass an informed judgment concerning what Heisenberg says in this book about the common-sense and Cartesian concepts of material and mental substances. This is the case because his conclusion concerning Descartes results from his generalization of the principle of complementarity beyond physics, first, to the relation between common-sense biological concepts and mathematical physical concepts and, second, to the body-mind problem. The result of this generalization is that the Cartesian theory of mental substances comes off very much better in this book, as does the concept of substance generally, than is the case in any other book on the philosophy of contemporary physics which this writer knows.

* Margenau, *op. cit.*, p. 422.

Whitehead, for example, concluded that contemporary science and philosophy find no place for, and have no need of, the concept of substance. Neutral monists such as Lord Russell and logical positivists such as Professor Carnap agree.

Generally speaking, Heisenberg argues that there is no compelling reason to throw away any of the common-sense concepts of either biology or mathematical physics, after one knows the refined concepts that lead to the complete clarification of the problems in atomic physics. Because the latter clarification is complete, it is relevant only to a very limited range of problems within science and cannot enable us to avoid using many concepts at other places that would not stand critical analysis of the type carried out in quantum theory. Since the ideal of complete clarification cannot be achieved—and it is important that we should not be deceived about this point—one may indulge in the usage of common-sense concepts if it is done with sufficient care and caution. In this respect, certainly, complementarity is a very useful scientific concept.

In any event, two things seem clear and make what Heisenberg says on these matters exceedingly important. First, the principle of complementarity and the present validity of the Cartesian and common-sense concepts of body and mind stand and fall together. Second, it may be that both these notions are merely convenient stepladders which should now be, or must eventually be, thrown away. Even so, in the case of the theory of mind at least, the stepladder will have to remain until by its use we find the more linguistically exact and empirically satisfactory theory that will permit us to throw the Cartesian language away. To be sure, piecemeal theories of mind which do not appeal to the notion of substance now exist, but none of their authors, unless it be Whitehead, has shown how the langauge of this piecemeal theory can be brought into commensurate and compatible relationship with the scientific language of the other facts of human knowledge. It is likely, therefore, that anyone, whether he be a professional physicist or philosopher or the lay reader, who may think he knows better than Heisenberg on these important matters, runs the grave risk of supposing he has a scientific theory of mind in its relation to body, when in fact this is not the case.

Up to this point we have directed attention, with but two exceptions, to what the philosophy of contemporary physics has to say about the object of scientific knowledge *qua* object, independent of its relation to the scientist as knower. In short, we have been concerned with its ontology. This philosophy also has its epistemological component. This component falls into three parts: (1) The relation between (a) the directly observed data given to the physicist as inductive knower in his observations or his experiments and (b) the speculatively proposed, indirectly verified, axiomatically constructed postulates of his theory. The latter term (b) defines the objects of scientific knowledge *qua* object and, hence, gives the ontology. The relation between (a) and (b) defines one factor in the epistemology. (2) The role of the concept of probability in the theory of errors, by means of which the physicist defines the criterion for judging how far his experimental findings can depart, due to errors of human experimentation, from the deduced consequences of the postulates of the theory and still be regarded as confirming the theory. (3) The effect of the experiment being performed upon the object being known. What Heisenberg says about the first and second of these three epistemological factors in contemporary physics has already been emphasized in this Introduction. It remains to direct the reader's attention to what he says about item (3).

In modern physical theory, previous to quantum mechanics, (3) played no role whatever. Hence, the epistemology of modern physics was then completely specified by (1) and (2) alone. In quantum mechanics, however, (3) (as well as (1) and (2)) becomes very important. The very act of observing alters the object being observed when its quantum numbers are small.

From this fact Heisenberg draws a very important conclusion concerning the relation between the object, the observing physicist, and the rest of the universe. This conclusion can be appreciated if attention is directed to the following key points. It may be recalled that in some of the definitions of mechanical causality given earlier in this Introduction, the qualifying words "for an isolated system" were added; elsewhere it was implicit. This qualifying condition can be satisfied in principle in Newtonian and Einsteinian mechanics, and also in practice by making more and more careful observations and refinements in one's experimental instruments. The introduction of the concept of probability into the definition of state of the object of scientific knowledge in quantum mechanics rules out, however, *in principle,* and not merely in practice due to the imperfections of human observation and instruments, the satisfying of the condition that the object of the physicist's knowledge is an isolated system. Heisenberg shows also that the including of the experimental apparatus and even of the eye of the observing scientist in the physical system which is the object of the knower's knowledge does not help, since, if quantum mechanics be correct, the states of all objects have to be defined in principle by recourse to the concept of probability. Consequently, only if the whole universe is included in the object of scientific knowledge can the qualifying condition "for an isolated system" be satisfied for even the weaker form of mechanical causation. Clearly, the philosophy of contemporary physics is shown by this book to be as novel in its epistemology as it is in its ontology. Indeed, it is from the originality of its ontology—the consistent unification of potentiality and mechanical causality in its weaker form—that the novelty of its epistemology arises.

Unquestionably, one other thing is clear. An analysis of the specific experimentally verified theories of modern physics with respect to what they say about the object of human knowledge and its relation to the human knower exhibits a very rich and complex ontological and epistemological philosophy which is an essential part of the scientific theory and method itself. Hence, physics is neither epistemologically nor ontologically neutral. Deny any one of the epistemological assumptions of the physicist's theory and there is no scientific method for testing whether what the theory says about the physical object is true, in the sense of being empirically confirmed. Deny any one of the ontological assumptions and there is not enough content in the axiomatically constructed mathematical postulates of the physicist's theory to permit the deduction of the experimental facts which it is introduced to predict, co-ordinate consistently and explain. Hence, to the extent that experimental physicists assure us that their theory of contemporary physics is indirectly and experimentally verified, they *ipso facto* assure us that its rich and complex ontological and epistemological philosophy is verified also.

When such empirically verified philosophy of the true in the natural sciences is identified with the criterion of the good and the just in the humanities and the social sciences, one has natural-law ethics and jurisprudence. In other words, one has a scientifically meaningful cognitive criterion and method for judging both the verbal, personal and social norms of the positive law and the living ethos embodied in the customs, habits and traditional cultural institutions of the *de facto* peoples and cultures of the world. It is the coming together of this new philosophy of physics with the respective philosophies of culture of mankind that is the major event in today's and tomorrow's world. At this point, the philosophy of physics in this book and its important reference to the social consequences of physics come together.

The chapters of this book have been read as Gifford Lectures at the University of St. Andrews during the winter term 1955–1956. According to the will of their founder the Gifford Lectures should "freely discuss all questions about man's conceptions of God or the Infinite, their origin, nature, and truth, whether he can have any such conceptions, whether God is under any or what limitations and so on." The lectures of Heisenberg do not attempt to reach these most general and most difficult problems. But they try to go far beyond the limited scope of a special science into the wide field of those general human problems that have been raised by the enormous recent development and the far-reaching practical applications of natural science.

I.

An Old and a New Tradition

WHEN one speaks today of modern physics, the first thought is of atomic weapons. Everybody realizes the enormous influence of these weapons on the political structure of our present world and is willing to admit that the influence of physics on the general situation is greater than it ever has been before. But is the political aspect of modern physics really the most important one? When the world has adjusted itself in its political structure to the new technical possibilities, what then will remain of the influence of modern physics?

To answer these questions, one has to remember that every tool carries with it the spirit by which it has been created. Since every nation and every political group has to be interested in the new weapons in some way irrespective of the location and of the cultural tradition of this group, the spirit of modern physics will penetrate into the minds of many people and will connect itself in different ways with the older traditions. What will be the outcome of this impact of a special branch of modern science on different powerful old traditions? In those parts of the world in which modern science has been developed the primary interest has been directed for a long time toward practical activity, industry and engineering combined with a rational analysis of the outer and inner conditions for such activity. Such people will find it rather easy to cope with the new ideas since they have had time for a slow and gradual adjustment to the modern scientific methods of thinking. In other parts of the world these ideas would be confronted with the religious and philosophical foundations of the native culture. Since it is true that the results of modern physics do touch such fundamental concepts as reality, space and time, the confrontation may lead to entirely new developments which cannot yet be foreseen. One characteristic feature of this meeting between modern science and the older methods of thinking will be its complete internationality. In this exchange of thoughts the one side, the old tradition, will be different in the different parts of the world, but the other side will be the same everywhere and therefore the results of this exchange will be spread over all areas in which the discussions take place.

For such reasons it may not be an unimportant task to try to discuss these ideas of modern physics in a not too technical language, to study their philosophical consequences, and to compare them with some of the older traditions.

The best way to enter into the problems of modern physics may be by a historical description of the development of quantum theory. It is true that quantum theory is only a small sector of atomic physics and atomic physics again is only a very small sector of modern science. Still it is in quantum theory that the most fundamental changes with respect to the concept of reality have taken place, and in quantum theory in its final form the new ideas of atomic physics are concentrated and crystallized. The enormous and extremely complicated experimental equipment needed for research in nuclear physics shows another very impressive aspect of this part of modern science. But with regard to the experimental technique nuclear physics represents the extreme extension of a method of research which has determined the growth of modern science ever since Huyghens or Volta or Faraday. In a similar sense the discouraging mathematical complication of some parts of quantum theory may be said to represent the extreme consequence of the methods of Newton or Gauss or Maxwell. But the change in the concept of reality manifesting itself in quantum theory is not simply a continuation of the past; it seems to be a real break in the structure of modern science. Therefore, the first of the following chapters will be devoted to the study of the historical development of quantum theory.

II.

The History of Quantum Theory

THE origin of quantum theory is connected with a well-known phenomenon, which did not belong to the central parts of atomic physics. Any piece of matter when it is heated starts to glow, gets red hot and white hot at higher temperatures. The color does not depend much on the surface of the material, and for a black body it depends solely on the temperature. Therefore, the radiation emitted by such a black body at high temperatures is a suitable object for physical research; it is a simple phenomenon that should find a simple explanation in terms of the known laws for radiation and heat. The attempt made at the end of the nineteenth century by Lord Rayleigh and Jeans failed, however, and revealed serious difficulties. It would not be possible to describe these difficulties here in simple terms. It must be sufficient to state that the application of the known laws did not lead to sensible results. When Planck, in 1895, entered this line of research he tried to turn the problem from radiation to the radiating atom. This turning did not remove any of the difficulties inherent in the problem, but it simplified the interpretation of the empirical facts. It was just at this time, during the summer of 1900, that Curlbaum and Rubens in Berlin had made very accurate new measurements of the spectrum of heat radiation. When Planck heard of these results he tried to represent them by simple mathematical formulas which looked plausible from his research on the general connection between heat and radiation. One day Planck and Rubens met for tea in Planck's home and compared Rubens' latest results with a new formula suggested by Planck. The comparison showed a complete agreement. This was the discovery of Planck's law of heat radiation.

It was at the same time the beginning of intense theoretical work for Planck. What was the correct physical interpretation of the new formula? Since Planck could, from his earlier work, translate his formula easily into a statement about the radiating atom (the so-called oscillator), he must soon have found that his formula looked as if the oscillator could only contain discrete quanta of energy—a result that was so different from anything known in classical physics that he certainly must have refused to believe it in the beginning. But in a period of most intensive work during the summer of 1900 he finally convinced himself that there was no way of escaping from this conclusion. It was told by Planck's son that his father spoke to him about his new ideas on a long walk through the Grunewald, the wood in the suburbs of Berlin. On this walk he explained that he felt he had possibly made a discovery of the first rank, comparable perhaps only to the discoveries of Newton. So Planck must have realized at this time that his formula had touched the foundations of our description of nature, and that these foundations would one day start to move from their traditional present location toward a new and as yet unknown position of stability. Planck, who was conservative in his whole outlook, did not like this consequence at all, but he published his quantum hypothesis in December of 1900.

The idea that energy could be emitted or absorbed only in discrete energy quanta was so new that it could not be fitted into the traditional framework of physics. An attempt by Planck to reconcile his new hypothesis with the older laws of radiation failed in the essential points. It took five years until the next step could be made in the new direction.

This time it was the young Albert Einstein, a revolutionary genius among the physicists, who was not afraid to go further away from the old concepts. There were two problems in which he could make use of the new ideas. One was the so-called photoelectric effect, the emission of electrons from metals under the influence of light. The experiments, especially those of Lenard, had shown that the energy of the emitted electrons did not depend on the intensity of the light, but only on its color or, more precisely, on its frequency. This could not be understood on the basis of the traditional theory of radiation. Einstein could explain the observations by interpreting Planck's hypothesis as saying that light consists of quanta of energy traveling through space. The energy of one light quantum should, in agreement with Planck's assumptions, be equal to the frequency of the light multiplied by Planck's constant.

The other problem was the specific heat of solid bodies. The traditional theory led to values for the specific heat which fitted the observations at higher temperatures but disagreed with them at low ones. Again Einstein was able to show that one could understand this behavior by applying the quantum hypothesis to the elastic vibrations of the atoms in the solid body. These two results marked a very important advance, since they revealed the presence of Planck's quantum of action—as his constant is called among the physicists—in several phenomena, which had nothing immediately to do with heat radiation. They revealed at the same time the deeply revolutionary character of the new hypothesis, since the first of them led to a description of light completely different from the traditional wave picture. Light could either be interpreted as consisting of electromagnetic waves, according to Maxwell's theory, or as consisting of light quanta, energy packets traveling through space with high velocity. But could it be both? Einstein knew, of course, that the well-known phenomena of diffraction and interference can be explained only on the basis of the wave picture. He was not able to dispute the complete contradiction between this wave picture and the idea of the light quanta; nor did he even attempt to remove the inconsistency of this interpretation. He simply took the contradiction as something which would probably be understood only much later.

In the meantime the experiments of Becquerel, Curie and Rutherford had led to some clarification concerning the structure of the atom. In 1911 Rutherford's observations on the interaction of α-rays penetrating through matter resulted in his famous atomic model. The atom is pictured as consisting of a nucleus, which is positively charged and contains nearly the total mass of the atom, and electrons, which circle around the nucleus like the planets circle around the sun. The chemical bond between atoms of different elements is explained as an interaction between the outer electrons of the neighboring atoms; it has not directly to do with the atomic nucleus. The nucleus determines the chemical behavior of the atom through its charge which in turn fixes the number of electrons in the neutral atom. Initially this model of the atom could not explain the most characteristic feature of the atom, its enormous stability. No planetary system following the laws of Newton's mechanics would ever go back to its original configuration after a collision with another such system. But an atom of the element carbon, for instance, will still remain a carbon atom after any collision or interaction in chemical binding.

The explanation for this unusual stability was given by Bohr in 1913, through the application of Planck's quantum hypothesis. If the atom can change its energy only by discrete energy quanta, this must mean that the atom can exist only in discrete stationary states, the lowest of which is the normal state of the atom. Therefore, after any kind of interaction the atom will finally always fall back into its normal state.

By this application of quantum theory to the atomic model, Bohr could not only explain the stability of the atom but also, in some simple cases, give a theoretical interpretation of the line spectra emitted by the atoms after the excitation through electric discharge or heat. His theory rested upon a combination of classical mechanics for the motion of the electrons with quantum conditions, which were imposed upon the classical motions for defining the discrete stationary states of the system. A consistent mathematical formulation for those conditions was later given by Sommerfeld. Bohr was well aware of the fact that the quantum conditions spoil in some way the consistency of Newtonian mechanics. In the simple case of the hydrogen atom one could calculate from Bohr's theory the frequencies of the light emitted by the atom, and the agreement with the observations was perfect. Yet these frequencies were different from the orbital frequencies and their harmonics of the electrons circling around the nucleus, and this fact showed at once that the theory was still full of contradictions. But it contained an essential part of the truth. It did explain qualitatively the chemical behavior of the atoms and their line spectra; the existence of the discrete stationary states was verified by the experiments of Franck and Hertz, Stern and Gerlach.

Bohr's theory had opened up a new line of research. The great amount of experimental material collected by spectroscopy through several decades was now available for information about the strange quantum laws governing the motions of the electrons in the atom. The many experiments of chemistry could be used for the same purpose. It was from this time on that the physicists learned to ask the right questions; and asking the right question is frequently more than halfway to the solution of the problem.

What were these questions? Practically all of them had to do with the strange apparent contradictions between the results of different experiments. How could it be that the same radiation that produces interference patterns, and therefore must consist of waves, also produces the photoelectric effect, and therefore must consist of moving particles? How could it be that the frequency of the orbital motion of the electron in the atom does not show up in the frequency of the emitted radiation? Does this mean that there is no orbital motion? But if the idea of orbital motion should in incorrect, what happens to the electrons inside the atom? One can see the electrons move through a cloud chamber, and sometimes they are knocked out of an atom; why should they not also move within the atom? It is true that they might be at rest in the normal state of the atom, the state of lowest energy. But there are many states of higher energy, where the electronic shell has an angular momentum. There the electrons cannot possibly be at rest. One could add a number of similar examples. Again and again one found that the attempt to describe atomic events in the traditional terms of physics led to contradictions.

Gradually, during the early twenties, the physicists became accustomed to these difficulties, they acquired a certain vague knowledge about where trouble would occur, and they learned to avoid contradictions. They knew which description of an atomic event would be the correct one for the special experiment under discussion. This was not sufficient to form a consistent general picture of what happens in a quantum process, but it changed the minds of the physicists in such a way that they somehow got into the spirit of quantum theory. Therefore, even some time before one had a consistent formulation of quantum theory one knew more or less what would be the result of any experiment.

One frequently discussed what one called ideal experiments. Such experiments were designed to answer a very critical question irrespective of whether or not they could actually be carried out. Of course it was important that it should be possible in principle to carry out the experiment, but the technique might be extremely complicated. These ideal experiments could be very useful in clarifying certain problems. If there was no agreement among the physicists about the result of such an ideal experiment, it was frequently possible to find a similar but simpler experiment that could be carried out, so that the experimental answer contributed essentially to the clarification of quantum theory.

The strangest experience of those years was that the paradoxes of quantum theory did not disappear during this process of clarification; on the contrary, they became even more marked and more exciting. There was, for instance, the experiment of Compton on the scattering of X-rays. From earlier experiments on the interference of scattered light there could be no doubt that scattering takes place essentially in the following way: The incident light wave makes an electron in the beam vibrate in the frequency of the wave; the oscillating electron then emits a spherical wave with the same frequency and thereby produces the scattered light. However, Compton found in 1923 that the frequency of scattered X-rays was different from the frequency of the incident X-ray. This change of frequency could be formally understood by assuming that scattering is to be described as collision of a light quantum with an electron. The energy of the light quantum is changed during the collision; and since the frequency times Planck's constant should be the energy of the light quantum, the frequency also should be changed. But what happens in this interpretation of the light wave? The two experiments—one on the interference of scattered light and the other on the change of frequency of the scattered light—seemed to contradict each other without any possibility of compromise.

By this time many physicists were convinced that these apparent contradictions belonged to the intrinsic structure of atomic physics. Therefore, in 1924 de Broglie in France tried to extend the dualism between wave description and particle description to the elementary particles of matter, primarily to the electrons. He showed that a certain matter wave could "correspond" to a moving electron, just as a light wave corresponds to a moving light quantum. It was not clear at the time what the word "correspond" meant in this connection. But de Broglie suggested that the quantum condition in Bohr's theory should be interpreted as a statement about the matter waves. A wave circling around a nucleus can for geometrical reasons only be a stationary wave; and the perimeter of the orbit must be an integer multiple of the wave length. In this way de Broglie's idea connected the quantum condition, which always had been a foreign element in the mechanics of the electrons, with the dualism between waves and particles.

In Bohr's theory the discrepancy between the calculated orbital frequency of the electrons and the frequency of the emitted radiation had to be interpreted as a limitation to the concept of the electronic orbit. This concept had been somewhat doubtful from the beginning. For the higher orbits, however, the electrons should move at a large distance from the nucleus just as they do when one sees them moving through a cloud chamber. There one should speak about electronic orbits. It was therefore very satisfactory that for these higher orbits the frequencies of the emitted radiation approach the orbital frequency and its higher harmonics. Also Bohr had already suggested in his early papers that the intensities of the emitted spectral lines approach the intensities of the corresponding harmonics. This principle of correspondence had proved very useful for the approximative calculation of the intensities of spectral lines. In this way one had the impression that Bohr's theory gave a qualitative but not a quantitative description of what happens inside the atom; that some new feature of the behavior of matter was qualitatively expressed by the quantum conditions, which in turn were connected with the dualism between waves and particles.

The precise mathematical formulation of quantum theory finally emerged from two different developments. The one started from Bohr's principle of correspondence. One had to give up the concept of the electronic orbit but still had to maintain it in the limit of high quantum numbers, i.e., for the large orbits. In this latter case the emitted radiation, by means of its frequencies and intensities, gives a picture of the electronic orbit; it represents what the mathematicians call a Fourier expansion of the orbit. The idea suggested itself that one should write down the mechanical laws not as equations for the positions and velocities of the electrons but as equations for the frequencies and amplitudes of their Fourier expansion. Starting from such equations and changing them very little one could hope to come to relations for those quantities which correspond to the frequencies and intensities of the emitted radiation, even for the small orbits and the ground state of the atom. This plan could actually be carried out; in the summer of 1925 it led to a mathematical formalism called matrix mechanics or, more generally, quantum mechanics. The equations of motion in Newtonian mechanics were replaced by similar equations between matrices; it was a strange experience to find that many of the old results of Newtonian mechanics, like conservation of energy, etc., could be derived also in the new scheme. Later the investigations of Born, Jordan and Dirac showed that the matrices representing position and momentum of the electron do not commute. This latter fact demonstrated clearly the essential difference between quantum mechanics and classical mechanics.

The other development followed de Broglie's idea of matter waves. Schrödinger tried to set up a wave equation for de Broglie's stationary waves around the nucleus. Early in 1926 he succeeded in deriving the energy values of the stationary states of the hydrogen atom as "Eigenvalues" of his wave equation and could give a more general prescription for transforming a given set of classical equations of motion into a corresponding wave equation in a space of many dimensions. Later he was able to prove that his formalism of wave mechanics was mathematically equivalent to the earlier formalism of quantum mechanics.

Thus one finally had a consistent mathematical formalism,

which could be defined in two equivalent ways starting either from relations between matrices or from wave equations. This formalism gave the correct energy values for the hydrogen atom; it took less than one year to show that it was also successful for the helium atom and the more complicated problems of the heavier atoms. But in what sense did the new formalism describe the atom? The paradoxes of the dualism between wave picture and particle picture were not solved; they were hidden somehow in the mathematical scheme.

A first and very interesting step toward a real understanding of quantum theory was taken by Bohr, Kramers and Slater in 1924. These authors tried to solve the apparent contradiction between the wave picture and the particle picture by the concept of the probability wave. The electromagnetic waves were interpreted not as "real" waves but as probability waves, the intensity of which determines in every point the probability for the absorption (or induced emission) of a light quantum by an atom at this point. This idea led to the conclusion that the laws of conservation of energy and momentum need not be true for the single event, that they are only statistical laws and are true only in the statistical average. This conclusion was not correct, however, and the connections between the wave aspect and the particle aspect of radiation were still more complicated.

But the paper of Bohr, Kramers and Slater revealed one essential feature of the correct interpretation of quantum theory. This concept of the probability wave was something entirely new in theoretical physics since Newton. Probability in mathematics or in statistical mechanics means a statement about our degree of knowledge of the actual situation. In throwing dice we do not know the fine details of the motion of our hands which determine the fall of the dice and therefore we say that the probability for throwing a special number is just one in six. The probability wave of Bohr, Kramers, Slater, however, meant more than that; it meant a tendency for something. It was a quantitative version of the old concept of "potentia" in Aristotelian philosophy. It introduced something standing in the middle between the idea of an event and the actual event, a strange kind of physical reality just in the middle between possibility and reality.

Later when the mathematical framework of quantum theory was fixed, Born took up this idea of the probability wave and gave a clear definition of the mathematical quantity in the formalism, which was to be interpreted as the probability wave. It was not a three-dimensional wave like elastic or radio waves, but a wave in the many-dimensional configuration space, and therefore a rather abstract mathematical quantity.

Even at this time, in the summer of 1926, it was not clear in every case how the mathematical formalism should be used to describe a given experimental situation. One knew how to describe the stationary states of an atom, but one did not know how to describe a much simpler event—as for instance an electron moving through a cloud chamber.

When Schrödinger in that summer had shown that his formalism of wave mechanics was mathematically equivalent to quantum mechanics he tried for some time to abandon the idea of quanta and "quantum jumps" altogether and to replace the electrons in the atoms simply by his three-dimensional matter waves. He was inspired to this attempt by his result, that the energy levels of the hydrogen atom in his theory seemed to be simply the eigenfrequencies of the stationary matter waves. Therefore, he thought it was a mistake to call them energies; they were just frequencies. But in the discussions which took place in the autumn of 1926 in Copenhagen between Bohr and Schrödinger and the Copenhagen group of physicists it soon became apparent that such an interpretation would not even be sufficient to explain Planck's formula of heat radiation.

During the months following these discussions an intensive study of all questions concerning the interpretation of quantum theory in Copenhagen finally led to a complete and, as many physicists believe, satisfactory clarification of the situation. But it was not a solution which one could easily accept. I remember discussions with Bohr which went through many hours till very late at night and ended almost in despair; and when at the end of the discussion I went alone for a walk in the neighboring park I repeated to myself again and again the question: Can nature possibly be as absurd as it seemed to us in these atomic experiments?

The final solution was approached in two different ways. The one was a turning around of the question. Instead of asking: How can one in the known mathematical scheme express a given experimental situation? the other question was put: Is it true, perhaps, that only such experimental situations can arise in nature as can be expressed in the mathematical formalism? The assumption that this was actually true led to limitations in the use of those concepts that had been the basis of classical physics since Newton. One could speak of the position and of the velocity of an electron as in Newtonian mechanics and one could observe and measure these quantities. But one could not fix both quantities simultaneously with an arbitrarily high accuracy. Actually the product of these two inaccuracies turned out to be not less than Planck's constant divided by the mass of the particle. Similar relations could be formulated for other experimental situations. They are usually called relations of uncertainty or principle of indeterminacy. One had learned that the old concepts fit nature only inaccurately.

The other way of approach was Bohr's concept of complementarity. Schrödinger had described the atom as a system not of a nucleus and electrons but of a nucleus and matter waves. This picture of the matter waves certainly also contained an element of truth. Bohr considered the two pictures—particle picture and wave picture—as two complementary descriptions of the same reality. Any of these descriptions can be only partially true, there must be limitations to the use of the particle concept as well as of the wave concept, else one could not avoid contradictions. If one takes into account those limitations which can be expressed by the uncertainty relations, the contradictions disappear.

In this way since the spring of 1927 one has had a consistent interpretation of quantum theory, which is frequently called the "Copenhagen interpretation." This interpretation received its crucial test in the autumn of 1927 at the Solvay conference in Brussels. Those experiments which had always led to the worst paradoxes were again and again discussed in all details, especially by Einstein. New ideal experiments were invented to trace any possible inconsistency of the theory, but the theory was shown to be consistent and seemed to fit the experiments as far as one could see.

The details of this Copenhagen interpretation will be the subject of the next chapter. It should be emphasized at this point that it has taken more than a quarter of a century to get from the first idea of the existence of energy quanta to a real understanding of the quantum theoretical laws. This indicates the great change that had to take place in the fundamental concepts concerning reality before one could understand the new situation.

III.

The Copenhagen Interpretation of Quantum Theory

THE Copenhagen interpretation of quantum theory starts from a paradox. Any experiment in physics, whether it refers to the phenomena of daily life or to atomic events, is to be described in the terms of classical physics. The concepts of classical physics form the language by which we describe the arrangement of our experiments and state the results. We cannot and should not replace these concepts by any others. Still the application of these concepts is limited by the relations of uncertainty. We must keep in mind this limited range of applicability of the classical concepts while using them, but we cannot and should not try to improve them.

For a better understanding of this paradox it is useful to compare the procedure for the theoretical interpretation of an experiment in classical physics and in quantum theory. In Newton's mechanics, for instance, we may start by measuring the position and the velocity of the planet whose motion we are going to study. The result of the observation is translated into mathematics by deriving numbers for the co-ordinates and the momenta of the planet from the observation. Then the equations of motion are used to derive from these values of the co-ordinates and momenta at a given time the values of these co-ordinates or any other properties of the system at a later time, and in this way the astronomer can predict the properties of the system at a later time. He can, for instance, predict the exact time for an eclipse of the moon.

In quantum theory the procedure is slightly different. We could for instance be interested in the motion of an electron through a cloud chamber and could determine by some kind of observation the initial position and velocity of the electron. But this determination will not be accurate; it will at least contain the inaccuracies following from the uncertainty relations and will probably contain still larger errors due to the difficulty of the experiment. It is the first of these inaccuracies which allows us to translate the result of the observation into the mathematical scheme of quantum theory. A probability function is written down which represents the experimental situation at the time of the measurement, including even the possible errors of the measurement.

This probability function represents a mixture of two things, partly a fact and partly our knowledge of a fact. It represents a fact in so far as it assigns at the initial time the probability unity (i.e., complete certainty) to the initial situation: the electron moving with the observed velocity at the observed position; "observed" means observed within the accuracy of the experiment. It represents our knowledge in so far as another observer could perhaps know the position of the electron more accurately. The error in the experiment does—at least to some extent—not represent a property of the electron but a deficiency in our knowledge of the electron. Also this deficiency of knowledge is expressed in the probability function.

In classical physics one should in a careful investigation also consider the error of the observation. As a result one would get a probability distribution for the initial values of the co-ordinates and velocities and therefore something very similar to the probability function in quantum mechanics. Only the necessary uncertainty due to the uncertainty relations is lacking in classical physics.

When the probability function in quantum theory has been determined at the initial time from the observation, one can from the laws of quantum theory calculate the probability function at any later time and can thereby determine the probability for a measurement giving a specified value of the measured quantity. We can, for instance, predict the probability for finding the electron at a later time at a given point in the cloud chamber. It should be emphasized, however, that the probability function does not in itself represent a course of events in the course of time. It represents a tendency for events and our knowledge of events. The probability function can be connected with reality only if one essential condition is fulfilled: if a new measurement is made to determine a certain property of the system. Only then does the probability function allow us to calculate the probable result of the new measurement. The result of the measurement again will be stated in terms of classical physics.

Therefore, the theoretical interpretation of an experiment requires three distinct steps: (1) the translation of the initial experimental situation into a probability function; (2) the following up of this function in the course of time; (3) the statement of a new measurement to be made of the system, the result of which can then be calculated from the probability function. For the first step the fulfillment of the uncertainty relations is a necessary condition. The second step cannot be described in terms of the classical concepts; there is no description of what happens to the system between the initial observation and the next measurement. It is only in the third step that we change over again from the "possible" to the "actual."

Let us illustrate these three steps in a simple ideal experiment. It has been said that the atom consists of a nucleus and electrons moving around the nucleus; it has also been stated that the concept of an electronic orbit is doubtful. One could argue that it should at least in principle be possible to observe the electron in its orbit. One should simply look at the atom through a microscope of a very high resolving power, then one would see the electron moving in its orbit. Such a high resolving power could to be sure not be obtained by a microscope using ordinary light, since the inaccuracy of the measurement of the position can never be smaller than the wave length of the light. But a microscope using γ-rays with a wave length smaller than the size of the atom would do. Such a microscrope has not yet been constructed but that should not prevent us from discussing the ideal experiment.

Is the first step, the translation of the result of the observation into a probability function, possible? It is possible only if the uncertainty relation is fulfilled after the observation. The position of the electron will be known with an accuracy given by the wave length of the γ-ray. The electron may have been practically at rest before the observation. But in the act of observation at least one light quantum of the γ-ray must have passed the microscope and must first have been deflected by the electron. Therefore, the electron has been pushed by the light quantum, it has changed its momentum and its velocity, and one can show that the uncertainty of this change is just big enough to guarantee the validity of the uncertainty relations. Therefore, there is no difficulty with the first step.

At the same time one can easily see that there is no way of observing the orbit of the electron around the nucleus. The second step shows a wave pocket moving not around the nucleus but away from the atom, because the first light quantum will have knocked the electron out from the atom. The momentum of light quantum of the γ-ray is much bigger than the original momentum of the electron if the wave length of the γ-ray is much smaller than the size of the atom. Therefore, the first light quantum is sufficient to knock the electron out of the atom and one can never observe more than one point in the orbit of the electron; therefore, there is no orbit in the ordinary sense. The next observation—the third step—will show the electron on its path from the atom. Quite generally there is no way of describing what happens between two consecutive observations. It is of course tempting to say that the electron must have been somewhere between the two observations and that therefore the electron must have described some kind of path or orbit even if it may be impossible to know which path. This would be a reasonable argument in classical physics. But in quantum theory it would be a misuse of the language which, as we will see later, cannot be justified. We can leave it open for the moment, whether this warning is a statement about the way in which we should talk about atomic events or a statement about the events themselves, whether it refers to epistemology or to ontology. In any case we have to be very cautious about the wording of any statement concerning the behavior of atomic particles.

Actually we need not speak of particles at all. For many experiments it is more convenient to speak of matter waves; for instance, of stationary matter waves around the atomic nucleus.

Such a description would directly contradict the other description if one does not pay attention to the limitations given by the uncertainty relations. Through the limitations the contradiction is avoided. The use of "matter waves" is convenient, for example, when dealing with the radiation emitted by the atom. By means of its frequencies and intensities the radiation gives information about the oscillating charge distribution in the atom, and there the wave picture comes much nearer to the truth than the particle picture. Therefore, Bohr advocated the use of both pictures, which he called "complementary" to each other. The two pictures are of course mutually exclusive, because a certain thing cannot at the same time be a particle (i.e., substance confined to a very small volume) and a wave (i.e., a field spread out over a large space), but the two complement each other. By playing with both pictures, by going from the one picture to the other and back again, we finally get the right impression of the strange kind of reality behind our atomic experiments. Bohr uses the concept of "complementarity" at several places in the interpretation of quantum theory. The knowledge of the position of a particle is complementary to the knowledge of its velocity or momentum. If we know the one with high accuracy we cannot know the other with high accuracy; still we must know both for determining the behavior of the system. The space-time description of the atomic events is complementary to their deterministic description. The probability function obeys an equation of motion as the co-ordinates did in Newtonian mechanics; its change in the course of time is completely determined by the quantum mechanical equation, but it does not allow a description in space and time. The observation, on the other hand, enforces the description in space and time but breaks the determined continuity of the probability function by changing our knowledge of the system.

Generally the dualism between two different descriptions of the same reality is no longer a difficulty since we know from the mathematical formulation of the theory that contradictions cannot arise. The dualism between the two complementary pictures—waves and particles—is also clearly brought out in the flexibility of the mathematical scheme. The formalism is normally written to resemble Newtonian mechanics, with equations of motion for the co-ordinates and the momenta of the particles. But by a simple transformation it can be rewritten to resemble a wave equation for an ordinary three-dimensional matter wave. Therefore, this possibility of playing with different complementary pictures has its analogy in the different transformations of the mathematical scheme; it does not lead to any difficulties in the Copenhagen interpretation of quantum theory.

A real difficulty in the understanding of this interpretation arises, however, when one asks the famous question: But what happens "really" in an atomic event? It has been said before that the mechanism and the results of an observation can always be stated in terms of the classical concepts. But what one deduces from an observation is a probability function, a mathematical expression that combines statements about possibilities or tendencies with statements about our knowledge of facts. So we cannot completely objectify the result of an observation, we cannot describe what "happens" between this observation and the next. This looks as if we had introduced an element of subjectivism into the theory, as if we meant to say: what happens depends on our way of observing it or on the fact that we observe it. Before discussing this problem of subjectivism it is necessary to explain quite clearly why one would get into hopeless difficulties if one tried to describe what happens between two consecutive observations.

For this purpose it is convenient to discuss the following ideal experiment: We assume that a small source of monochromatic light radiates toward a black screen with two small holes in it. The diameter of the holes may be not much bigger than the wave length of the light, but their distance will be very much bigger. At some distance behind the screen a photographic plate registers the incident light. If one describes this experiment in terms of the wave picture, one says that the primary wave penetrates through the two holes; there will be secondary spherical waves starting from the holes that interfere with one another, and the interference will produce a pattern of varying intensity on the photographic plate.

The blackening of the photographic plate is a quantum process, a chemical reaction produced by single light quanta. Therefore, it must also be possible to describe the experiment in terms of light quanta. If it would be permissible to say what happens to the single light quantum between its emission from the light source and its absorption in the photographic plate, one could argue as follows: The single light quantum can come through the first hole or through the second one. If it goes through the first hole and is scattered there, its probability for being absorbed at a certain point of the photographic plate cannot depend upon whether the second hole is closed or open. The probability distribution on the plate will be the same as if only the first hole was open. If the experiment is repeated many times and one takes together all cases in which the light quantum has gone through the first hole, the blackening of the plate due to these cases will correspond to this probability distribution. If one considers only those light quanta that go through the second hole, the blackening should correspond to a probability distribution derived from the assumption that only the second hole is open. The total blackening, therefore, should just be the sum of the blackenings in the two cases; in other words, there should be no interference pattern. But we know this is not correct, and the experiment will show the interference pattern. Therefore, the statement that any light quantum must have gone *either* through the first *or* through the second hole is problematic and leads to contradictions. This example shows clearly that the concept of the probability function does not allow a description of what happens between two observations. Any attempt to find such a description would lead to contradictions; this must mean that the term "happens" is restricted to the observation.

Now, this is a very strange result, since it seems to indicate that the observation plays a decisive role in the event and that the reality varies, depending upon whether we observe it or not. To make this point clearer we have to analyze the process of observation more closely.

To begin with, it is important to remember that in natural science we are not interested in the universe as a whole, including ourselves, but we direct our attention to some part of the universe and make that the object of our studies. In atomic physics this part is usually a very small object, an atomic particle or a group of such particles, sometimes much larger—the size does not matter; but it is important that a large part of the universe, including ourselves, does *not* belong to the object.

Now, the theoretical interpretation of an experiment starts with the two steps that have been discussed. In the first step we have to describe the arrangement of the experiment, eventually combined with a first observation, in terms of classical physics and translate this description into a probability function. This probability function follows the laws of quantum theory, and its change in the course of time, which is continuous, can be calculated from the initial conditions; this is the second step. The probability function combines objective and subjective elements. It contains statements about possibilities or better tendencies ("potentia" in Aristotelian philosophy), and these statements are completely objective, they do not depend on any observer; and it contains statements about our knowledge of the system, which of course are subjective in so far as they may be different for different observers. In ideal cases the subjective element in the probability function may be practically negligible as compared with the objective one. The physicists then speak of a "pure case."

When we now come to the next observation, the result of which should be predicted from the theory, it is very important to realize that our object has to be in contact with the other part of the world, namely, the experimental arrangement, the measuring rod, etc., before or at least at the moment of observation. This means that the equation of motion for the probability function does now contain the influence of the interaction with the measuring device. This influence introduces a new element of uncertainty, since the measuring device is necessarily described in the terms of classical physics; such a description contains all the uncertainties concerning the microscopic structure of the device which we know from thermodynamics, and since the device is connected with the rest of the world, it contains in fact the uncertainties of the microscopic structure of the whole world. These uncertainties may be called objective in so far as they are simply a consequence of the description in the terms of classical physics and do not depend on any observer. They may be called subjective in so far as they refer to our incomplete knowledge of the world.

After this interaction has taken place, the probability function contains the objective element of tendency and the subjective element of incomplete knowledge, even if it has been a "pure case" before. It is for this reason that the result of the observation cannot generally be predicted with certainty; what can be predicted is the probability of a certain result of the observation, and this statement about the probability can be checked by repeating the experiment many times. The probability function does—unlike the common procedure in Newtonian mechanics—not describe a certain event but, at least during the process of observation, a whole ensemble of possible events.

The observation itself changes the probability function discontinuously; it selects of all possible events the actual one that has taken place. Since through the observation our knowledge of the system has changed discontinuously, its mathematical representation also has undergone the discontinuous change and we speak of a "quantum jump." When the old adage "Natura non facit saltus" is used as a basis for criticism of quantum theory, we can reply that certainly our knowledge can change suddenly and that this fact justifies the use of the term "quantum jump."

Therefore, the transition from the "possible" to the "actual" takes place during the act of observation. If we want to describe what happens in an atomic event, we have to realize that the word "happens" can apply only to the observation, not to the state of affairs between two observations. It applies to the physical, not the psychical act of observation, and we may say that the transition from the "possible" to the "actual" takes place as soon as the interaction of the object with the measuring device, and thereby with the rest of the world, has come into play; it is not connected with the act of registration of the result by the mind of the observer. The discontinuous change in the probability function, however, takes place with the act of registration, because it is the discontinuous change of our knowledge in the instant of registration that has its image in the discontinuous change of the probability function.

To what extent, then, have we finally come to an objective description of the world, especially of the atomic world? In classical physics science started from the belief—or should one say from the illusion?—that we could describe the world or at least parts of the world without any reference to ourselves. This is actually possible to a large extent. We know that the city of London exists whether we see it or not. It may be said that classical physics is just that idealization in which we can speak about parts of the world without any reference to ourselves. Its success has led to the general ideal of an objective description of the world. Objectivity has become the first criterion for the value of any scientific result. Does the Copenhagen interpretation of quantum theory still comply with this ideal? One may perhaps say that quantum theory corresponds to this ideal as far as possible. Certainly quantum theory does not contain genuine subjective features, it does not introduce the mind of the physicist as a part of the atomic event. But it starts from the division of the world into the "object" and the rest of the world, and from the fact that at least for the rest of the world we use the classical concepts in our description. This division is arbitrary and historically a direct consequence of our scientific method; the use of the classical concepts is finally a consequence of the general human way of thinking. But this is already a reference to ourselves and in so far our description is not completely objective.

It has been stated in the beginning that the Copenhagen interpretation of quantum theory starts with a paradox. It starts from the fact that we describe our experiments in the terms of classical physics and at the same time from the knowledge that these concepts do not fit nature accurately. The tension between these two starting points is the root of the statistical character of quantum theory. Therefore, it has sometimes been suggested that one should depart from the classical concepts altogether and that a radical change in the concepts used for describing the experiments might possibly lead back to a nonstatical, completely objective description of nature.

This suggestion, however, rests upon a misunderstanding. The concepts of classical physics are just a refinement of the concepts of daily life and are an essential part of the language which forms the basis of all natural science. Our actual situation in science is such that we *do* use the classical concepts for the description of the experiments, and it was the problem of quantum theory to find theoretical interpretation of the experiments on this basis. There is no use in discussing what could be done if we were other beings than we are. At this point we have to realize, as von Weizsäcker has put it, that "Nature is earlier than man, but man is earlier than natural science." The first part of the sentence justifies classical physics, with its ideal of complete objectivity. The second part tells us why we cannot escape the paradox of quantum theory, namely, the necessity of using the classical concepts.

We have to add some comments on the actual procedure in the quantum-theoretical interpretation of atomic events. It has been said that we always start with a division of the world into an object, which we are going to study, and the rest of the world, and that this division is to some extent arbitrary. It should indeed not make any difference in the final result if we, e.g., add some part of the measuring device or the whole device to the object and apply the laws of quantum theory to this more complicated object. It can be shown that such an alteration of the theoretical treatment would not alter the predictions concerning a given experiment. This follows mathematically from the fact that the laws of quantum theory are for the phenomena in which Planck's constant can be considered as a very small quantity, approximately identical with the classical laws. But it would be a mistake to believe that this application of the quantum-theoretical laws to the measuring device could help to avoid the fundamental paradox of quantum theory.

The measuring device deserves this name only if it is in close contact with the rest of the world, if there is an interaction between the device and the observer. Therefore, the uncertainty with respect to the microscopic behavior of the world will enter into the quantum-theoretical system here just as well as in the first interpretation. If the measuring device would be isolated from the rest of the world, it would be neither a measuring device nor could it be described in the terms of classical physics at all.

With regard to this situation Bohr has emphasized that it is more realistic to state that the division into the object and the rest of the world is not arbitrary. Our actual situation in research work in atomic physics is usually this: we wish to understand a certain phenomenon, we wish to recognize how this phenomenon follows from the general laws of nature. Therefore, that part of matter or radiation which takes part in the phenomenon is the natural "object" in the theoretical treatment and should be separated in this respect from the tools used to study the phenomenon. This again emphasizes a subjective element in the description of atomic events, since the measuring device has been constructed by the observer, and we have to remember that what we observe is not nature in itself but nature exposed to our method of questioning. Our scientific work in physics consists in asking questions about nature in the language that we possess and trying to get an answer from experiment by the means that are at our disposal. In this way quantum theory reminds us, as Bohr has put it, of the old wisdom that when searching for harmony in life one must never forget that in the drama of existence we are ourselves both players and spectators. It is understandable that in our scientific relation to nature our own activity becomes very important when we have to deal with parts of nature into which we can penetrate only by using the most elaborate tools.

IV.

Quantum Theory and the Roots of Atomic Science

THE concept of the atom goes back much further than the beginning of modern science in the seventeenth century; it has its origin in ancient Greek philosophy and was in that early period the central concept of materialism taught by Leucippus and Democritus. On the other hand, the modern interpretation of atomic events has very little resemblance to genuine materialistic philosophy; in fact, one may say that atomic physics has turned science away from the materialistic trend it had during the nineteenth century. It is therefore interesting to compare the development of Greek philosophy toward the concept of the atom with the present position of this concept in modern physics.

The idea of the smallest, indivisible ultimate building blocks of matter first came up in connection with the elaboration of the concepts of Matter, Being and Becoming which characterized the first epoch of Greek philosophy. This period started in the sixth century B.C. with Thales, the founder of the Milesian school, to whom Aristotle ascribes the statement: "Water is the material cause of all things." This statement, strange as it looks to us, expresses, as Nietzsche has pointed out, three fundamental ideas of philosophy. First, the question as to the material cause of all things; second, the demand that this question be answered in conformity with reason, without resort to myths or mysticism; third, the postulate that ultimately it must be possible to reduce everything to one principle. Thales' statement was the first expression of the idea of a fundamental substance, of which all other things were transient forms. The word "substance" in this connection was certainly in that age not interpreted in the purely material sense which we frequently ascribe to it today. Life was connected with or inherent in this "substance" and Aristotle ascribes to Thales also the statement: All things are full of gods. Still the question was put as to the material cause of all things and it is not difficult to imagine that Thales took his view primarily from meteorological considerations. Of all things we know water can take the most various shapes; it can in the winter take the form of ice and snow, it can change into vapor, and it can form the clouds. It seems to turn into earth where the rivers form their delta, and it can spring from the earth. Water is the condition for life. Therefore, if there was such a fundamental substance, it was natural to think of water first.

The idea of the fundamental substance was then carried further by Anaximander, who was a pupil of Thales and lived in the same town. Anaximander denied the fundamental substance to be water or any of the known substances. He taught that the primary substance was infinite, eternal and ageless and that it encompassed the world. This primary substance is transformed into the various substances with which we are familiar. Theophrastus quotes from Anaximander: "Into that from which things take their rise they pass away once more, as is ordained, for they make reparation and satisfaction to one another for their injustice according to the ordering of time." In this philosophy the antithesis of Being and Becoming plays the fundamental role. The primary substance, infinite and ageless, the undifferentiated Being, degenerates into the various forms which lead to endless struggles. The process of Becoming is considered as a sort of debasement of the infinite Being—a disintegration into the struggle ultimately expiated by a return into that which is without shape or character. The struggle which is meant here is the opposition between hot and cold, fire and water, wet and dry, etc. The temporary victory of the one over the other is the injustice for which they finally make reparation in the ordering of time. According to Anaximander, there is "eternal motion," the creation and passing away of worlds from infinity to infinity.

It may be interesting to notice at this point that the problem—whether the primary substance can be one of the known substances or must be something essentially different—occurs in a somewhat different form in the most modern part of atomic physics. The physicists today try to find a fundamental law of motion for matter from which all elementary particles and their properties can be derived mathematically. This fundamental equation of motion may refer either to waves of a known type, to proton and meson waves, or to waves of an essentially different character which have nothing to do with any of the known waves or elementary particles. In the first case it would mean that all other elementary particles can be reduced in some way to a few sorts of "fundamental" elementary particles; actually theoretical physics has during the past two decades mostly followed this line of research. In the second case all different elementary particles could be reduced to some universal substance which we may call energy or matter, but none of the different particles could be preferred to the others as being more fundamental. The latter view of course corresponds to the doctrine of Anaximander, and I am convinced that in modern physics this view is the correct one. But let us return to Greek philosophy.

The third of the Milesian philosophers, Anaximenes, an associate of Anaximander, taught that air was the primary substance. "Just as our soul, being air, holds us together, so do breath and air encompass the whole world." Anaximenes introduced into the Milesian philosophy the idea that the process of condensation or rarefaction causes the change of the primary substance into the other substances. The condensation of water vapor into clouds was an obvious example, and of course the difference between water vapor and air was not known at that time.

In the philosophy of Heraclitus of Ephesus the concept of Becoming occupies the foremost place. He regarded that which moves, the fire, as the basic element. The difficulty, to reconcile the idea of one fundamental principle with the infinite variety of phenomena, is solved for him by recognizing that the strife of the opposites is really a kind of harmony. For Heraclitus the world is at once one and many, it is just "the opposite tension" of the opposites that constitutes the unity of the One. He says: "We must know that war is common to all and strife is justice, and that all things come into being and pass away through strife."

Looking back to the development of Greek philosophy up to this point one realizes that it has been borne from the beginning to this stage by the tension between the One and the Many. For our senses the world consists of an infinite variety of things and events, colors and sounds. But in order to understand it we have to introduce some kind of order, and order means to recognize what is equal, it means some sort of unity. From this springs the belief that there is one fundamental principle, and at the same time the difficulty to derive from it the infinite variety of things. That there should be a material cause for all things was a natural starting point since the world consists of matter. But when one carried the idea of fundamental unity to the extreme one came to that infinite and eternal undifferentiated Being which, whether material or not, cannot in itself explain the infinite variety of things. This leads to the antithesis of Being and Becoming and finally to the solution of Heraclitus, that the change itself is the fundamental principle; the "imperishable change, that renovates the world," as the poets have called it. But the change in itself is not a material cause and therefore is represented in the philosophy of Heraclitus by the fire as the basic element, which is both matter and a moving force.

We may remark at this point that modern physics is in some way extremely near to the doctrines of Heraclitus. If we replace the word "fire" by the word "energy" we can almost repeat his statements word for word from our modern point of view. Energy is in fact the substance from which all elementary particles, all atoms and therefore all things are made, and energy is that which moves. Energy is a substance, since its total amount does not change, and the elementary particles can actually be made from this substance as is seen in many experiments on the creation of elementary particles. Energy can be changed into motion, into heat, into light and into tension. Energy may be called the fundamental cause for all change in the world. But this comparison of Greek philosophy with the ideas of modern science will be discussed later.

Greek philosophy returned for some time to the concept of the One in the teachings of Parmenides, who lived in Elea in the south of Italy. His most important contribution to Greek thinking was, perhaps, that he introduced a purely logical argument into metaphysics. "One cannot know what is not—that is impossible—nor utter it; for it is the same thing that can be thought and that can be." Therefore, only the One is, and there is no becoming or passing away. Parmenides denied the existence of empty space for logical reasons. Since all change requires empty space, as he assumed, he dismissed change as an illusion.

But philosophy could not rest for long on this paradox. Empedocles, from the south coast of Sicily, changed for the first time from monism to a kind of pluralism. To avoid the difficulty that one primary substance cannot explain the variety of things and events, he assumed four basic elements, earth, water, air and fire. The elements are mixed together and separated by the action of Love and Strife. Therefore, these latter two, which are in many ways treated as corporeal like the other four elements, are responsible for the imperishable change. Empedocles describes the formation of the world in the following picture: First, there is the infinite Sphere of the One, as in the philosophy of Parmenides. But in the primary substance all the four "roots" are mixed together by Love. Then, when Love is passing out and Strife coming in, the elements are partially separated and partially combined. After that the elements are completely separated and Love is outside the World. Finally, Love is bringing the elements together again and Strife is passing out, so that we return to the original Sphere.

This doctrine of Empedocles represents a very definite turning toward a more materialistic view in Greek philosophy. The four elements are not so much fundamental principles as real material substances. Here for the first time the idea is expressed that the mixture and separation of a few substances, which are fundamentally different, explains the infinite variety of things and events. Pluralism never appeals to those who are wont to think in fundamental principles. But it is a reasonable kind of compromise, which avoids the difficulty of monism and allows the establishment of some order.

The next step toward the concept of the atom was made by Anaxagoras, who was a contemporary of Empedocles. He lived in Athens about thirty years, probably in the first half of the fifth century B.C. Anaxagoras stresses the idea of the mixture, the assumption that all change is caused by mixture and separation. He assumes an infinite variety of infinitely small "seeds," of which all things are composed. The seeds do not refer to the four elements of Empedocles, there are innumerably many different seeds. But the seeds are mixed together and separated again and in this way all change is brought about. The doctrine of Anaxagoras allows for the first time a geometrical interpretation of the term "mixture": Since he speaks of the infinitely small seeds, their mixture can be pictured as the mixture between two kinds of sand of different colors. The seeds may change in number and in relative position. Anaxagoras assumes that all seeds are in everything, only the proportion may change from one thing to another. He says: "All things will be in everything; nor is it possible for them to be apart, but all things have a portion of everything." The universe of Anaxagoras is set in motion not by Love and Strife, like that of Empedocles, but by "Nous," which we may translate as "Mind."

From this philosophy it was only one step to the concept of the atom, and this step occurred with Leucippus and Democritus of Abdera. The antithesis of Being and Not-being in the philosophy of Parmenides is here secularized into the antithesis of the "Full" and the "Void." Being is not only One, it can be repeated an infinite number of times. This is the atom, the indivisible smallest unit of matter. The atom is eternal and indestructible, but it has a finite size. Motion is made possible through the empty space between the atoms. Thus for the first time in history there was voiced the idea of the existence of smallest ultimate particles—we would say of elementary particles, as the fundamental building blocks of matter.

According to this new concept of the atom, matter did not consist only of the "Full," but also of the "Void," of the empty space in which the atoms move. The logical objection of Parmenides against the Void, that not-being cannot exist, was simply ignored to comply with experience. From our modern point of view we would say that the empty space between the atoms in the philosophy of Democritus was not nothing; it was the carrier for geometry and kinematics, making possible the various arrangements and movements of atoms. But the possibility of empty space has always been a controversial problem in philosophy. In the theory of general relativity the answer is given that geometry is produced by matter or matter by geometry. This answer corresponds more closely to the view held by many philosophers that space is defined by the extension of matter. But Democritus clearly departs from this view, to make change and motion possible.

The atoms of Democritus were all of the same substance, which had the property of being, but had different sizes and different shapes. They were pictured therefore as divisible in a mathematical but not in a physical sense. The atoms could move and could occupy different positions in space. But they had no other physical properties. They had neither color nor smell nor taste. The properties of matter which we perceive by our senses were supposed to be produced by the movements and positions of the atoms in space. Just as both tragedy and comedy can be written by using the same letters of the alphabet, the vast variety of events in this world can be realized by the same atoms through their different arrangements and movements. Geometry and kinematics, which were made possible by the void, proved to be still more important in some way than pure being. Democritus is quoted to have said: "A thing merely appears to have color, it merely appears to be sweet or bitter. Only atoms and empty space have a real existence."

The atoms in the philosophy of Leucippus do not move merely by chance. Leucippus seems to have believed in complete determinism, since he is known to have said: "Naught happens for nothing, but everything from a ground and of necessity." The atomists did not give any reason for the original motion of the atoms, which just shows that they thought of a causal description of the atomic motion; causality can only explain later events by earlier events, but it can never explain the beginning.

The basic ideas of atomic theory were taken over and modified, in part, by later Greek philosophers. For the sake of comparison with modern atomic physics it is important to mention the explanation of matter given by Plato in his dialogue *Timaeus*. Plato was not an atomist; on the contrary, Diogenes Laertius reported that Plato disliked Democritus so much that he wished all his books to be burned. But Plato combined ideas that were near to atomism with the doctrines of the Pythagorean school and the teachings of Empedocles.

The Pythagorean school was an offshoot of Orphism, which goes back to the worship of Dionysus. Here has been established the connection between religion and mathematics which ever since has exerted the strongest influence on human thought. The Pythagoreans seem to have been the first to realize the creative force inherent in mathematical formulations. Their discovery that two strings sound in harmony if their lengths are in a simple ratio demonstrated how much mathematics can mean for the understanding of natural phenomena. For the Pythagoreans it was not so much a question of understanding. For them the simple mathematical ratio between the length of the strings *created* the harmony in sound. There was also much mysticism in the doctrines of the Pythagorean school which for us is difficult to understand. But by making mathematics a part of their religion they touched an essential point in the development of human thought. I may quote a statement by Bertrand Russell about Pythagoras: "I do not know of any other man who has been as influential as he was in the sphere of thought."

Plato knew of the discovery of the regular solids made by the Pythagoreans and of the possibility of combining them with the elements of Empedocles. He compared the smallest parts of the element earth with the cube, of air with the octahedron, of fire with the tetrahedron, and of water with the icosahedron. There is no element that corresponds to the dodecahedron; here Plato only says "there was yet a fifth combination which God used in the delineation of the universe."

If the regular solids, which represent the four elements, can be compared with the atoms at all, it is made clear by Plato that they are not indivisible. Plato constructs the regular solids from two basic triangles, the equilateral and the isosceles triangles, which are put together to form the surface of the solids. Therefore, the elements can (at least partly) be transformed into each other. The regular solids can be taken apart into their triangles and new regular solids can be formed of them. For instance, one tetrahedron and two octahedra can be taken apart into twenty equilateral triangles, which can be recombined to give one icosahedron. That means: one atom of fire and two atoms of air can be combined to give one atom of water. But the fundamental triangles cannot be considered as matter, since they have no extension in space. It is only when the triangles are put together to form a regular solid that a unit of matter is created. The smallest parts of matter are not the fundamental Beings, as in the philosophy of Democritus, but are mathematical forms. Here it is quite evident that the form is more important than the substance of which it is the form.

After this short survey of Greek philosophy up to the formation of the concept of the atom we may come back to modern physics and ask how our modern views on the atom and on quantum theory compare with this ancient development. Historically the word "atom" in modern physics and chemistry was referred to the wrong object, during the revival of science in the seventeenth century, since the smallest particles belonging to what is called a chemical element are still rather complicated systems of smaller units. These smaller units are nowadays called elementary particles, and it is obvious that if anything in modern physics should be compared with the atoms of Democritus it should be the elementary particles like proton, neutron, electron, meson.

Democritus was well aware of the fact that if the atoms should, by their motion and arrangement, *explain* the properties of matter—color, smell, taste—they cannot themselves have these properties. Therefore, he has deprived the atom of these qualities and his atom is thus a rather abstract piece of matter. But Democritus has left to the atom the quality of "being," of extension in space, of shape and motion. He has left these qualities because it would have been difficult to speak about the atom at all if such qualities had been taken away from it. On the other hand, this implies that his concept of the atom cannot explain geometry, extension in space or existence, because it cannot reduce them to something more fundamental. The modern view of the elementary particle with regard to this point seems more consistent and more radical. Let us discuss the question: What *is* an elementary particle? We say, for instance, simply "a neutron" but we can give no well-defined picture and what we mean by the word. We can use several pictures and describe it once as a particle, once as a wave or as a wave packet. But we know that none of these descriptions is accurate. Certainly the neutron has no color, no smell, no taste. In this respect it resembles the atom of Greek philosophy. But even the other qualities are taken from the elementary particle, at least to some extent; the concepts of geometry and kinematics, like shape or motion in space, cannot be applied to it consistently. If one wants to give an accurate description of the elementary particle—and here the emphasis is on the word "accurate"—the only thing which can be written down as description is a probability function. But then one sees that not even the quality of being (if that may be called a "quality") belongs to what is described. It is a possibility for being or a tendency for being. Therefore, the elementary particle of modern physics is still far more abstract than the atom of the Greeks, and it is by this very property more consistent as a clue for explaining the behavior of matter.

In the philosophy of Democritus all atoms consist of the same substance if the word "substance" is to be applied here at all. The elementary particles in modern physics carry a mass in the same limited sense in which they have other properties. Since mass and energy are, according to the theory of relativity, essentially the same concepts, we may say that all elementary particles consist of energy. This could be interpreted as defining energy as the primary substance of the world. It has indeed the essential property belonging to the term "substance," that it is conserved. Therefore, it has been mentioned before that the views of modern physics are in this respect very close to those of Heraclitus if one interprets his element fire as meaning energy. Energy is in fact that which moves; it may be called the primary cause of all change, and energy can be transformed into matter or heat or light. The strife between opposites in the philosophy of Heraclitus can be found in the strife between two different forms of energy.

In the philosophy of Democritus the atoms are eternal and indestructible units of matter, they can never be transformed into each other. With regard to this question modern physics takes a definite stand against the materialism of Democritus and for Plato and the Pythagoreans. The elementary particles are certainly not eternal and indestructible units of matter, they can actually be transformed into each other. As a matter of fact, if two such particles, moving through space with a very high kinetic energy, collide, then many new elementary particles may be created from the available energy and the old particles may have disappeared in the collision. Such events have been frequently observed and offer the best proof that all particles are made of the same substance: energy. But the resemblance of the modern views to those of Plato and the Pythagoreans can be carried somewhat further. The elementary particles in Plato's *Timaeus* are finally not substance but mathematical forms. "All things are numbers" is a sentence attributed to Pythagoras. The only mathematical forms available at that time were such geometric forms as the regular solids or the triangles which form their surface. In modern quantum theory there can be no doubt that the elementary particles will finally also be mathematical forms, but of a much more complicated nature. The Greek philosophers thought of static forms and found them in the regular solids. Modern science, however, has from its beginning in the sixteenth and seventeenth centuries started from the dynamic problem. The constant element in physics since Newton is not a configuration or a geometrical form, but a dynamic law. The equation of motion holds at all times, it is in this sense eternal, whereas the geometrical forms, like the orbits, are changing. Therefore, the mathematical forms that represent the elementary particles will be solutions of some eternal law of motion for matter. Actually this is a problem which has not yet been solved. The fundamental law of motion for matter is not yet known and therefore it is not yet possible to derive mathematically the properties of the elementary particles from such a law. But theoretical physics in its present state seems to be not very far from this goal and we can at least say what kind of law we have to expect. The final equation of motion for matter will probably be some quantized nonlinear wave equation for a wave field of operators that simply represents matter, not any specified kind of waves or particles. This wave equation will probably be equivalent to rather complicated sets of integral equations, which have "Eigenvalues" and "Eigensolutions," as the physicists call it. These Eigensolutions will finally represent the elementary particles; they are the mathematical forms which shall replace the regular solids of the Pythagoreans. We might mention here that these "Eigensolutions" will follow from the fundamental equation for matter by much the same mathematical process by which the harmonic vibrations of the Pythagorean string follow from the differential equation of the string. But, as has been said, these problems are not yet solved.

If we follow the Pythagorean line of thought we may hope

that the fundamental law of motion will turn out as a mathematically simple law, even if its evaluation with respect to the Eigenstates may be very complicated. It is difficult to give any good argument for this hope for simplicity—except the fact that it has hitherto always been possible to write the fundamental equations in physics in simple mathematical forms. This fact fits in with the Pythagorean religion, and many physicists share their belief in this respect, but no convincing argument has yet been given to show that it must be so.

We may add an argument at this point concerning a question which is frequently asked by laymen with respect to the concept of the elementary particle in modern physics: Why do the physicists claim that their elementary particles cannot be divided into smaller bits? The answer to this question clearly shows how much more abstract modern science is as compared to Greek philosophy. The argument runs like this: How could one divide an elementary particle? Certainly only by using extreme forces and very sharp tools. The only tools available are other elementary particles. Therefore, collisions between two elementary particles of extremely high energy would be the only processes by which the particles could eventually be divided. Actually they *can* be divided in such processes, sometimes into very many fragments; but the fragments are again elementary particles, not any smaller pieces of them, the masses of these fragments resulting from the very large kinetic energy of the two colliding particles. In other words, the transmutation of energy into matter makes it possible that the fragments of elementary particles are again the same elementary particles.

After this comparison of the modern views in atomic physics with Greek philosophy we have to add a warning, that this comparison should not be misunderstood. It may seem at first sight that the Greek philosophers have by some kind of ingenious intuition come to the same or very similar conclusions as we have in modern times only after several centuries of hard labor with experiments and mathematics. This interpretation of our comparison would, however, be a complete misunderstanding. There is an enormous difference between modern science and Greek philosophy, and that is just the empiristic attitude of modern science. Since the time of Galileo and Newton, modern science has been based upon a detailed study of nature and upon the postulate that only such statements should be made, as have been verified or at least can be verified by experiment. The idea that one could single out some events from nature by an experiment, in order to study the details and to find out what is the constant law in the continuous change, did not occur to the Greek philosophers. Therefore, modern science has from its beginning stood upon a much more modest, but at the same time much firmer, basis than ancient philosophy. Therefore, the statements of modern physics are in some way meant much more seriously than the statements of Greek philosophy. When Plato says, for instance, that the smallest particles of fire are tetrahedrons, it is not quite easy to see what he really means. Is the form of the tetrahedron only symbolically attached to the element fire, or do the smallest particles of fire mechanically act as rigid tetrahedrons or as elastic tetrahedrons, and by what force could they be separated into the equilateral triangles, etc.? Modern science would finally always ask: How can one decide experimentally that the atoms of fire are tetrahedrons and not perhaps cubes? Therefore, when modern science states that the proton is a certain solution of a fundamental equation of matter it means that we can from this solution deduce mathematically all possible properties of the proton and can check the correctness of the solution by experiments in every detail. This possibility of checking the correctness of a statement experimentally with very high precision and in any number of details gives an enormous weight to the statement that could not be attached to the statements of early Greek philosophy.

All the same, some statements of ancient philosophy are rather near to those of modern science. This simply shows how far one can get by combining the ordinary experience of nature that we have without doing experiments with the untiring effort to get some logical order into this experience to understand it from general principles.

V.

The Development of Philosophical Ideas Since Descartes in Comparison with the New Situation in Quantum Theory

IN THE two thousand years that followed the culmination of Greek science and culture in the fifth and fourth centuries B.C. the human mind was to a large extent occupied with problems of a different kind from those of the early period. In the first centuries of Greek culture the strongest impulse had come from the immediate reality of the world in which we live and which we perceive by our senses. This reality was full of life and there was no good reason to stress the distinction between matter and mind or between body and soul. But in the philosophy of Plato one already sees that another reality begins to become stronger. In the famous simile of the cave Plato compares men to prisoners in a cave who are bound and can look in only one direction. They have a fire behind them and see on a wall the shadows of themselves and of objects behind them. Since they see nothing but the shadows, they regard those shadows as real and are not aware of the objects. Finally one of the prisoners escapes and comes from the cave into the light of the sun. For the first time he sees real things and realizes that he had been deceived hitherto by the shadows. For the first time he knows the truth and thinks only with sorrow of his long life in the darkness. The real philosopher is the prisoner who has escaped from the cave into the light of truth, he is the one who possesses real knowledge. This immediate connection with truth or, we may in the Christian sense say, with God is the new reality that has begun to become stronger than the reality of the world as perceived by our senses. The immediate connection with God happens within the human soul, not in the world, and this was the problem that occupied human thought more than anything else in the two thousand years following Plato. In this period the eyes of the philosophers were directed toward the human soul and its relation to God, to the problems of ethics, and to the interpretation of the revelation but not to the outer world. It was only in the time of the Italian Renaissance that again a gradual change of the human mind could be seen, which resulted finally in a revival of the interest in nature.

The great development of natural science since the sixteenth and seventeenth centuries was preceded and accompanied by a development of philosophical ideas which were closely connected with the fundamental concepts of science. It may therefore be instructive to comment on these ideas from the position that has finally been reached by modern science in our time.

The first great philosopher of this new period of science was René Descartes who lived in the first half of the seventeenth century. Those of his ideas that are most important for the development of scientific thinking are contained in his *Discourse on Method*. On the basis of doubt and logical reasoning he tries to find a completely new and as he thinks solid ground for a philosophical system. He does not accept revelation as such a basis nor does he want to accept uncritically what is perceived by the senses. So he starts with his method of doubt. He casts his doubt upon that which our senses tell us about the results of our reasoning and finally he arrives at his famous sentence: "cogito ergo sum." I cannot doubt my existence since it follows from the fact that I am thinking. After establishing the existence of the I in this way he proceeds to prove the existence of God essentially on the lines of scholastic philosophy. Finally the existence of the world follows from the fact that God had given me a strong inclination to believe in the existence of the world, and it is simply impossible that God should have deceived me.

This basis of the philosophy of Descartes is radically different from that of the ancient Greek philosophers. Here the starting point is not a fundamental principle or substance, but the attempt of a fundamental knowledge. And Descartes realizes that what we know about our mind is more certain than what we know about the outer world. But already his starting point with the "triangle" God-World-I simplifies in a dangerous way the basis for further reasoning. The division between matter and mind or between soul and body, which had started in Plato's philosophy, is now complete. God is separated both from the I and from the world. God in fact is raised so high above the world and men that He finally appears in the philosophy of Descartes only as a common point of reference that establishes the relation between the I and the world.

While ancient Greek philosophy had tried to find order in the infinite variety of things and events by looking for some fundamental unifying principle, Descartes tries to establish the order through some fundamental division. But the three parts which result from the division lose some of their essence when any one part is considered as separated from the other two parts. If one uses the fundamental concepts of Descartes at all, it is essential that God is in the world and in the I and it is also essential that the I cannot be really separated from the world. Of course Descartes knew the undisputable necessity of the connection, but philosophy and natural science in the following period developed on the basis of the polarity between the "res cogitans" and the "res extensa," and natural science concentrated its interest on the "res extensa." The influence of the Cartesian division on human thought in the following centuries can hardly be overestimated, but it is just this division which we have to criticize later from the development of physics in our time.

Of course it would be wrong to say that Descartes, through his new method in philosophy, has given a new direction to human thought. What he actually did was to formulate for the first time a trend in human thinking that could already be seen during the Renaissance in Italy and in the Reformation. There was the revival of interest in mathematics which expressed an increasing influence of Platonic elements in philosophy, and the insistence on personal religion. The growing interest in mathematics favored a philosophical system that started from logical reasoning and tried by this method to arrive at some truth that was as certain as a mathematical conclusion. The insistence on personal religion separated the I and its relation to God from the world. The interest in the combination of empirical knowledge with mathematics as seen in the work of Galileo was perhaps partly due to the possibility of arriving in this way at some knowledge that could be kept apart completely from the theological disputes raised by the Reformation. This empirical knowledge could be formulated without speaking about God or about ourselves and favored the separation of the three fundamental concepts God-World-I or the separation between "res cogitans" and "res extensa." In this period there was in some cases an explicit agreement among the pioneers of empirical science that in their discussions the name of God or a fundamental cause should not be mentioned.

On the other hand, the difficulties of the separation could be clearly seen from the beginning. In the distinction, for instance, between the "res cogitans" and the "res extensa" Descartes was forced to put the animals entirely on the side of the "res extensa." Therefore, the animals and the plants were not essentially different from machines, their behavior was completely determined by material causes. But it has always seemed difficult to deny completely the existence of some kind of soul in the animals, and it seems to us that the older concept of soul for instance in the philosophy of Thomas Aquinas was more natural and less forced than the Cartesian concept of the "res cogitans," even if we are convinced that the laws of physics and chemistry are strictly valid in living organisms. One of the later consequences of this view of Descartes was that, if animals were simply considered as machines, it was difficult not to think the same about men. Since, on the other hand, the "res cogitans" and the "res extensa" were taken as completely different in their essence, it did not seem possible that they could act upon each other. Therefore, in order to preserve complete parallelism between the experiences of the mind and of the body, the mind also was in its activities completely determined by laws which corresponded to the laws of physics and chemistry. Here the question of the possibility of "free will" arose. Obviously this whole description is somewhat artificial and shows the grave defects of the Cartesian partition.

On the other hand in natural science the partition was for several centuries extremely successful. The mechanics of Newton and all the other parts of classical physics constructed after its model started from the assumption that one can describe the world without speaking about God or ourselves. This possibility soon seemed almost a necessary condition for natural science in general.

But at this point the situation changed to some extent through quantum theory and therefore we may now come to a comparison of Descartes's philosophical system with our present situation in modern physics. It has been pointed out before that in the Copenhagen interpretation of quantum theory we can indeed proceed without mentioning ourselves as individuals, but we cannot disregard the fact that natural science is formed by men. Natural science does not simply describe and explain nature; it is a part of the interplay between nature and ourselves; it describes nature as exposed to our method of questioning. This was a possibility of which Descartes could not have thought, but it makes the sharp separation between the world and the I impossible.

If one follows the great difficulty which even eminent scientists like Einstein had in understanding and accepting the Copenhagen interpretation of quantum theory, one can trace the roots of this difficulty to the Cartesian partition. This partition has penetrated deeply into the human mind during the three centuries following Descartes and it will take a long time for it to be replaced by a really different attitude toward the problem of reality.

The position to which the Cartesian partition has led with respect to the "res extensa" was what one may call metaphysical realism. The world, i.e., the extended things, "exist." This is to be distinguished from practical realism, and the different forms of realism may be described as follows: We "objectivate" a statement if we claim that its content does not depend on the conditions under which it can be verified. Practical realism assumes that there are statements that can be objectivated and that in fact the largest part of our experience in daily life consists of such statements. Dogmatic realism claims that there are no statements concerning the material world that cannot be objectivated. Practical realism has always been and will always be an essential part of natural science. Dogmatic realism, however, is, as we see it now, not a necessary condition for natural science. But it has in the past played a very important role in the development of science; actually the position of classical physics is that of dogmatic realism. It is only through quantum theory that we have learned that exact science is possible without the basis of dogmatic realism. When Einstein has criticized quantum theory he has done so from the basis of dogmatic realism. This is a very natural attitude. Every scientist who does research work feels that he is looking for something that is objectively true. His statements are not meant to depend upon the conditions under which they can be verified. Especially in physics the fact that we can explain nature by simple mathematical laws tells us that here we have met some genuine feature of reality, not something that we have—in any meaning of the word—invented ourselves. This is the situation which Einstein had in mind when he took dogmatic realism as the basis for natural science. But quantum theory is in itself an example for the possibility of explaining nature by means of simple mathematical laws without this basis. These laws may perhaps not seem quite simple when one compares them with Newtonian mechanics. But, judging from the enormous complexity of the phenomena which are to be explained (for instance, the line spectra of complicated atoms), the mathematical scheme of quantum theory is comparatively simple. Natural science is actually possible without the basis of dogmatic realism.

Metaphysical realism goes one step further than dogmatic realism by saying that "the things really exist." This is in fact what Descartes tried to prove by the argument that "God cannot have deceived us." The statement that the things really exist is different from the statement of dogmatic realism in so far as here the word "exists" occurs, which is also meant in the other statement "cogito ergo sum" . . . "I think, therefore I am." But it is difficult to see what is meant at this point that is not yet contained in the thesis of dogmatic realism; and this leads us to a general criticism of the statement "cogito ergo sum," which Descartes considered as the solid ground on which he could build his system. It is in fact true that this statement has the certainty of a mathematical conclusion, if the words "cogito" and "sum" are defined in the usual way or, to put it more cautiously and at the same time more critically, if the words are so defined that the statement follows. But this does not tell us anything about how far we can use the concepts of "thinking" and "being" in finding our way. It is finally in a very general sense always an empirical question how far our concepts can be applied.

The difficulty of metaphysical realism was felt soon after Descartes and became the starting point for the empiristic philosophy, for sensualism and positivism.

The three philosophers who can be taken as representatives for early empiristic philosophy are Locke, Berkeley and Hume. Locke holds, contrary to Descartes, that all knowledge is ultimately founded in experience. This experience may be sensation or perception of the operation of our own mind. Knowledge, so Locke states, is the perception of the agreement or disagreement of two ideas. The next step was taken by Berkeley. If actualy all our knowledge is derived from perception, there is no meaning in the statement that the things really exist; because if the perception is given it cannot possibly make any difference whether the things exist or do not exist. Therefore, to be perceived is identical with existence. This line of argument then was extended to an extreme skepticism by Hume, who denied induction and causation and thereby arrived at a conclusion which if taken seriously would destroy the basis of all empirical science.

The criticism of metaphysical realism which has been expressed in empiristic philosophy is certainly justified in so far as it is a warning against the naïve use of the term "existence." The positive statements of this philosophy can be criticized on similar lines. Our perceptions are not primarily bundles of colors or sounds; what we perceive is already perceived as something, the accent here being on the word "thing," and therefore it is doubtful whether we gain anything by taking the perceptions instead of the things as the ultimate elements of reality.

The underlying difficulty has been clearly recognized by modern positivism. This line of thought expresses criticism against the naïve use of certain terms like "thing," "perception," "existence" by the general postulate that the question whether a given sentence has any meaning at all should always be thoroughly and critically examined. This postulate and its underlying attitude are derived from mathematical logic. The procedure of natural science is pictured as an attachment of symbols to the phenomena. The symbols can, as in mathematics, be combined according to certain rules, and in this way statements about the phenomena can be represented by combinations of symbols. However, a combination of symbols that does not comply with the rules is not wrong but conveys no meaning.

The obvious difficulty in this argument is the lack of any general criterion as to when a sentence should be considered as meaningless. A definite decision is possible only when the sentence belongs to a closed system of concepts and axioms, which in the development of natural science will be rather the exception than the rule. In some cases the conjecture that a certain sentence is meaningless has historically led to important progress, for it opened the way to the establishment of new connections which would have been impossible if the sentence had a meaning. An example in quantum theory that has already been discussed is the sentence: "In which orbit does the electron move around the nucleus?" But generally the positivistic scheme taken from mathematical logic is too narrow in a description of nature which necessarily uses words and concepts that are only vaguely defined.

The philosophic thesis that all knowledge is ultimately founded in experience has in the end led to a postulate concerning the logical clarification of any statement about nature. Such a postulate may have seemed justified in the period of classical physics, but since quantum theory we have learned that it cannot be fulfilled. The words "position" and "velocity" of an electron, for instance, seemed perfectly well defined as to both their meaning and their possible connections, and in fact they were clearly defined concepts within the mathematical framework of Newtonian mechanics. But actually they were not well defined, as is seen from the relations of uncertainty. One may say that regarding their position in Newtonian mechanics they were well defined, but in their relation to nature they were not. This shows that we can never know beforehand which limitations will be put on the applicability of certain concepts by the extension of our knowledge into the remote parts of nature, into which we can only penetrate with the most elaborate tools. Therefore, in the process of penetration we are bound sometimes to use our concepts in a way which is not justified and which carries no meaning. Insistence on the postulate of complete logical clarification would make science impossible. We are reminded here by modern physics of the old wisdom that the one who insists on never uttering an error must remain silent.

A combination of those two lines of thought that started from Descartes, on the one side, and from Locke and Berkeley, on the other, was attempted in the philosophy of Kant, who was the founder of German idealism. That part of his work which is important in comparison with the results of modern physics is contained in *The Critique of Pure Reason.* He takes up the question whether knowledge is only founded in experience or can come from other sources, and he arrives at the conclusion that our knowledge is in part "a priori" and not inferred inductively from experience. Therefore, he distinguishes between "empirical" knowledge and knowledge that is "a priori." At the same time he distinguishes between "analytic" and "synthetic" propositions. Analytic propositions follow simply from logic, and their denial would lead to self-contradiction. Propositions that are not "analytic" are called "synthetic."

What is, according to Kant, the criterion for knowledge being "a priori"? Kant agrees that all knowledge starts with experience but he adds that it is not always derived from experience. It is true that experience teaches us that a certain thing has such or such properties, but it does not teach us that it could not be different. Therefore, if a proposition is thought together with its necessity it must be "a priori." Experience never gives to its judgments complete generality. For instance, the sentence "The sun rises every morning" means that we know no exception to this rule in the past and that we expect it to hold in future. But we can imagine exceptions to the rule. If a judgment is stated with complete generality, therefore, if it is impossible to imagine any exception, it must be "a priori." An analytic judgment is always "a priori"; even if a child learns arithmetic from playing with marbles, he need not later go back to experience to know that "two and two are four." Empirical knowledge, on the other hand, is synthetic.

But are synthetic judgments a priori possible? Kant tries to prove this by giving examples in which the above criteria seem to be fulfilled. Space and time are, he says, a priori forms of pure intuition. In the case of space he gives the following metaphysical arguments:

1. Space is not an empirical concept, abstracted from other experiences, for space is presupposed in referring sensations to something external, and external experience is only possible through the presentation of space.
2. Space is a necessary presentation a priori, which underlies all external perceptions; for we cannot imagine that there should be no space, although we can imagine that there should be nothing in space.
3. Space is not a discursive or general concept of the relations of things in general, for there is only one space, of which what we call "spaces" are parts, not instances.
4. Space is presented as an infinite given magnitude, which holds within itself all the parts of space; this relation is different from that of a concept to its instances, and therefore space is not a concept but a form of intuition.

These arguments shall not be discussed here. They are mentioned merely as examples for the general type of proof that Kant has in mind for the synthetic judgments a priori.

With regard to physics Kant took as a priori, besides space and time, the law of causality and the concept of substance. In a later stage of his work he tried to include the law of conservation of matter, the equality of "actio and reactio" and even the law of gravitation. No physicist would be willing to follow Kant here, if the term "a priori" is used in the absolute sense that was given to it by Kant. In mathematics Kant took Euclidean geometry as "a priori."

Before we compare these doctrines of Kant with the results of modern physics we must mention another part of his work, to which we will have to refer later. The disagreeable question whether "the things really exist," which had given rise to empiristic philosophy, occurred also in Kant's system. But Kant has not followed the line of Berkeley and Hume, though that would have been logically consistent. He kept the notion of the "thing-in-itself" as different from the percept, and in this way kept some connection with realism.

Coming now to the comparison of Kant's doctrines with modern physics, it looks in the first moment as though his central concept of the "synthetic judgments a priori" had been completely annihilated by the discoveries of our century. The theory of relativity has changed our views on space and time, it has in fact revealed entirely new features of space and time, of which nothing is seen in Kant's a priori forms of pure intuition. The law of causality is no longer applied in quantum theory and the law of conservation of matter is no longer true for the elementary particles. Obviously Kant could not have foreseen the new discoveries, but since he was convinced that his concepts would be "the basis of any future metaphysics that can be called science" it is interesting to see where his arguments have been wrong.

As example we take the law of causality. Kant says that whenever we observe an event we assume that there is a foregoing event from which the other event must follow according to some rule. This is, as Kant states, the basis of all scientific work. In this discussion it is not important whether or not we can always find the foregoing event from which the other one followed. Actually we can find it in many cases. But even if we cannot, nothing can prevent us from asking what this foregoing event might have been and to look for it. Therefore, the law of causality is reduced to the method of scientific research; it is the condition which makes science possible. Since we actually apply this method, the law of causality is "a priori" and is not derived from experience.

Is this true in atomic physics? Let us consider a radium atom, which can emit an α-particle. The time for the emission of the α-particle cannot be predicted. We can only say that in the average the emission will take place in about two thousand years. Therefore, when we observe the emission we do not actually look for a foregoing event from which the emission must according to a rule follow. Logically it would be quite possible to look for such a foregoing event, and we need not be discouraged by the fact that hitherto none has been found. But why has the scientific method actually changed in this very fundamental question since Kant?

Two possible answers can be given to that question. The one is: We have been convinced by experience that the laws of quantum theory are correct and, if they are, we know that a foregoing event as cause for the emission at a given time cannot be found. The other answer is: We know the foregoing event, but not quite accurately. We know the forces in the atomic nucleus that are responsible for the emission of the α-particle.

But this knowledge contains the uncertainty which is brought about by the interaction between the nucleus and the rest of the world. If we wanted to know why the α-particle was emitted at that particular time we would have to know the microscopic structure of the whole world including ourselves, and that is impossible. Therefore, Kant's arguments for the a priori character of the law of causality no longer apply.

A similar discussion could be given on the a priori character of space and time as forms of intuition. The result would be the same. The a priori concepts which Kant considered an undisputable truth are no longer contained in the scientific system of modern physics.

Still they form an essential part of this system in a somewhat different sense. In the discussion of the Copenhagen interpretation of quantum theory it has been emphasized that we use the classical concepts in describing our experimental equipment and more generally in describing that part of the world which does not belong to the object of the experiment. The use of these concepts, including space, time and causality, is in fact the condition for observing atomic events and is, in this sense of the word, "a priori." What Kant had not foreseen was that these a priori concepts can be the conditions for science and at the same time can have only a limited range of applicability. When we make an experiment we have to assume a causal chain of events that leads from the atomic event through the apparatus finally to the eye of the observer; if this causal chain was not assumed, nothing could be known about the atomic event. Still we must keep in mind that classical physics and causality have only a limited range of applicability. It was the fundamental paradox of quantum theory that could not be foreseen by Kant. Modern physics has changed Kant's statement about the possibility of synthetic judgments a priori from a metaphysical one into a practical one. The synthetic judgments a priori thereby have the character of a relative truth.

If one reinterprets the Kantian "a priori" in this way, there is no reason to consider the perceptions rather than the things as given. Just as in classical physics, we can speak about those events that are not observed in the same manner as about those that are observed. Therefore, practical realism is a natural part of the reinterpretation. Considering the Kantian "thing-in-itself" Kant had pointed out that we cannot conclude anything from the perception about the "thing-in-itself." This statement has, as Weizsäcker has noticed, its formal analogy in the fact that in spite of the use of the classical concepts in all the experiments a nonclassical behavior of the atomic objects is possible. The "thing-in-itself" is for the atomic physicist, if he uses this concept at all, finally a mathematical structure; but this structure is—contrary to Kant—indirectly deduced from experience.

In this reinterpretation the Kantian "a priori" is indirectly connected with experience in so far as it has been formed through the development of the human mind in a very distant past. Following this argument the biologist Lorentz has once compared the "a priori" concepts with forms of behavior that in animals are called "inherited or innate schemes." It is in fact quite plausible that for certain primitive animals space and time are different from what Kant calls our "pure intuition" of space and time. The latter may belong to the species "man," but not to the world as independent of men. But we are perhaps entering into too hypothetical discussions by following this biological comment on the "a priori." It was mentioned here merely as an example of how the term "relative truth" in connection with the Kantian "a priori" can possibly be interpreted.

Modern physics has been used here as an example or, we may say, as a model to check the results of some important philosophic systems of the past, which of course were meant to hold in a much wider field. What we have learned especially from the discussion of the philosophies of Descartes and Kant may perhaps be stated in the following way:

Any concepts or words which have been formed in the past through the interplay between the world and ourselves are not really sharply defined with respect to their meaning; that is to say, we do not know exactly how far they will help us in finding our way in the world. We often know that they can be applied to a wide range of inner or outer experience, but we practically never know precisely the limits of their applicability. This is true even of the simplest and most general concepts like "existence" and "space and time." Therefore, it will never be possible by pure reason to arrive at some absolute truth.

The concepts may, however, be sharply defined with regard to their connections. This is actually the fact when the concepts become a part of a system of axioms and definitions which can be expressed consistently by a mathematical scheme. Such a group of connected concepts may be applicable to a wide field of experience and will help us to find our way in this field. But the limits of the applicability will in general not be known, at least not completely.

Even if we realize that the meaning of a concept is never defined with absolute precision, some concepts form an integral part of scientific methods, since they represent for the time being the final result of the development of human thought in the past, even in a very remote past; they may even be inherited and are in any case the indispensable tools for doing scientific work in our time. In this sense they can be practically a priori. But further limitations of their applicability may be found in the future.

VI.

The Relation of Quantum Theory to Other Parts of Natural Science

IT HAS been stated before that the concepts of natural science can sometimes be sharply defined with regard to their connections. This possibility was realized for the first time in Newton's *Principia* and it is just for that reason that Newton's work has exerted its enormous influence on the whole development of natural science in the following centuries. Newton begins his *Principia* with a group of definitions and axioms which are interconnected in such a way that they form what one may call a "closed system." Each concept can be represented by a mathematical symbol, and the connections between the different concepts are then represented by mathematical equations expressed by means of the symbols. The mathematical image of the system ensures that contradictions cannot occur in the system. In this way the possible motions of bodies under the influence of the acting forces are represented by the possible solutions of the equations. The system of definitions and axioms which can be written in a set of mathematical equations is considered as describing an eternal structure of nature, depending neither on a particular space nor on particular time.

The connection between the different concepts in the system is so close that one could generally not change any one of the concepts without destroying the whole system.

For this reason Newton's system was for a long time considered as final and the task set before the scientists of the following period seemed simply to be an expansion of Newton's mechanics into wider fields of experience. Actually physics did develop along these lines for about two centuries.

From the theory of the motion of mass points one could go over to the mechanics of solid bodies, to rotatory motions, and one could treat the continuous motions of a fluid or the vibrating motions of an elastic body. All these parts of mechanics or dynamics were gradually developed in close connection with the evolution of mathematics, especially of the differential calculus, and the results were checked by experiments. Acoustics and hydrodynamics became a part of mechanics. Another science, in which the application of Newton's mechanics was obvious, was astronomy. The improvements of the mathematical methods gradually led to more and more accurate determinations of the motions of the planets and of their mutual interactions. When the phenomena of electricity and magnetism were discovered, the electric or magnetic forces were compared to the gravitational forces and their actions upon the motion of the bodies could again be studied along the lines of Newtonian mechanics. Finally, in the nineteenth century, even the theory of heat could be reduced to mechanics by the assumption that heat really consists of a complicated statistical motion of the smallest parts of matter. By combining the concepts of the mathematical theory of probability with the concepts of Newtonian mechanics Clausius, Gibbs and Boltzmann were able to show that the fundamental laws in the theory of heat could be interpreted as statistical laws following from Newton's mechanics when applied to very complicated mechanical systems.

Up to this point the program set up by Newtonian mechanics had been carried out quite consistently and had led to the understanding of a wide field of experience. The first difficulty arose in the discussions on the electromagnetic field in the work of Faraday and Maxwell. In Newtonian mechanics the gravitational force had been considered as given, not as an object for further theoretical studies. In the work of Faraday and Maxwell, however, the field of force itself became the object of the investigation; the physicists wanted to know how this field of force varied as function of space and time. Therefore, they tried to set up equations of motion for the fields, not primarily for the bodies upon which the fields act. This change led back to a point of view which had been held by many scientists before Newton. An action could, so it seemed, be transferred from one body to another only when the two bodies touched each other; for instance, in a collision or through friction. Newton had introduced a very new and strange hypothesis by assuming a force that acted over a long distance. Now in the theory of the fields of force one could come back to the older idea, that action is transferred from one point to a neighboring point, only by describing the behavior of the fields in terms of differential equations. This proved actually to be possible, and therefore the description of the electromagnetic fields as given by Maxwell's equations seemed a satisfactory solution of the problem of force. Here one had really changed the program given by Newtonian mechanics. The axioms and definitions of Newton had referred to bodies and their motion; but with Maxwell the fields of force seemed to have acquired the same degree of reality as the bodies in Newton's theory. This view of course was not easily accepted; and to avoid such a change in the concept of reality it seemed plausible to compare the electromagnetic fields with the fields of elastic deformation or stress, the light waves of Maxwell's theory with the sound waves in elastic bodies. Therefore, many physicists believed that Maxwell's equations actually referred to the deformations of an elastic medium, which they called the ether; and this name was given merely to explain that the medium was so light and thin that it could penetrate into other matter and could not be seen or felt. This explanation was not too satisfactory, however, since it could not explain the complete absence of any longitudinal light waves.

Finally the theory of relativity, which will be discussed in the next chapter, showed in a conclusive way that the concept of the ether as a substance, to which Maxwell's equations refer, had to be abandoned. The arguments cannot be discussed at this point; but the result was that the fields had to be considered as an independent reality.

A further and still more startling result of the theory of special relativity was the discovery of new properties of space and time, actually of a relation between space and time that had not been known before and did not exist in Newtonian mechanics.

Under the impression of this completely new situation many physicists came to the following somewhat rash conclusion: Newtonian mechanics had finally been disproved. The primary reality is the field and not the body, and the structure of space and time is correctly described by the formulas of Lorentz and Einstein, and not by the axioms of Newton. The mechanics of Newton was a good approximation in many cases, but now it must be improved to give a more rigorous description of nature.

From the point of view which we have finally reached in

quantum theory such a statement would appear as a very poor description of the actual situation. First, it ignores the fact that most experiments by which fields are measured are based upon Newtonian mechanics and, second, that Newtonian mechanics cannot be improved; it can only be replaced by something essentially different!

The development of quantum theory has taught us that one should rather describe the situation in the following terms: Wherever the concepts of Newtonian mechanics can be used to describe events in nature, the laws formulated by Newton are strictly correct and cannot be improved. But the electromagnetic phenomena cannot adequately be described by the concepts of Newtonian mechanics. Therefore, the experiments on the electromagnetic fields and on light waves, together with their theoretical analysis by Maxwell, Lorentz and Einstein, have led to a new closed system of definitions and axioms and of concepts that can be represented by mathematical symbols, which is coherent in the same sense as the system of Newton's mechanics, but is essentially different from it.

Therefore, even the hopes which had accompanied the work of the scientists since Newton had to be changed. Apparently progress in science could not always be achieved by using the known laws of nature for explaining new phenomena. In some cases new phenomena that had been observed could only be understood by new concepts which were adapted to the new phenomena in the same way as Newton's concepts were to the mechanical events. These new concepts again could be connected in a closed system and represented by mathematical symbols. But if physics or, more generally, natural science proceeded in this way, the question arose: What is the relation between the different sets of concepts? If, for instance, the same concepts or words occur in two different sets and are defined differently with regard to their connection and mathematical representation, in what sense do the concepts represent reality?

This problem arose at once when the theory of special relativity had been discovered. The concepts of space and time belonged both to Newtonian mechanics and to the theory of relativity. But space and time in Newtonian mechanics were independent; in the theory of relativity they were connected by the Lorentz transformation. In this special case one could show that the statements of the theory of relativity approached those of Newtonian mechanics within the limit in which all velocities in the system are very small as compared with the velocity of light. From this one could conclude that the concepts of Newtonian mechanics could not be applied to events in which there occurred velocities comparable to the velocity of light. Thereby one had finally found an essential limitation of Newtonian mechanics which could not be seen from the coherent set of concepts nor from simple observations on mechanical systems.

Therefore, the relation between two different coherent sets of concepts always requires very careful investigation. Before we enter into a general discussion about the structure of any such closed and coherent set of concepts and about their possible relations we will give a brief description of those sets of concepts that have so far been defined in physics. One can distinguish four systems that have already attained their final form.

The first set, Newtonian mechanics, has already been discussed. It is suited for the description of all mechanical systems, of the motion of fluids and the elastic vibration of bodies; it comprises acoustics, statics, aerodynamics.

The second closed system of concepts was formed in the course of the nineteenth century in connection with the theory of heat. Though the theory of heat could finally be connected with mechanics through the development of statistical mechanics, it would not be realistic to consider it as a part of mechanics. Actually the phenomenological theory of heat uses a number of concepts that have no counterpart in other branches of physics, like: heat, specific heat, entropy, free energy, etc. If from this phenomenological description one goes over to a statistical interpretation, by considering heat as energy, distributed statistically among the very many degrees of freedom due to the atomic structure of matter, then heat is no more connected with mechanics than with electrodynamics or other parts of physics. The central concept of this interpretation is the concept of probability, closely connected with the concept of entropy in the phenomenological theory. Besides this concept the statistical theory of heat requires the concept of energy. But any coherent set of axioms and concepts in physics will necessarily contain the concepts of energy, momentum and angular momentum and the law that these quantities must under certain conditions be conserved. This follows if the coherent set is intended to describe certain features of nature that are correct at all times and everywhere; in other words, features that do not depend on space and time or, as the mathematicians put it, that are invariant under arbitrary translations in space and time, rotations in space and the Galileo—or Lorentz—transformation. Therefore, the theory of heat can be combined with any of the other closed systems of concepts.

The third closed system of concepts and axioms has its origin in the phenomena of electricity and magnetism and has reached its final form in the first decade of the twentieth century through the work of Lorentz, Einstein and Minkowski. It comprises electrodynamics, special relativity, optics, magnetism, and one may include the de Broglie theory of matter waves of all different sorts of elementary particles, but not the wave theory of Schrödinger.

Finally, the fourth coherent system is essentially the quantum theory as it has been described in the first two chapters. Its central concept is the probability function, or the "statistical matrix," as the mathematicians call it. It comprises quantum and wave mechanics, the theory of atomic spectra, chemistry, and the theory of other properties of matter like electric conductivity, ferromagnetism, etc.

The relations between these four sets of concepts can be indicated in the following way: The first set is contained in the third as the limiting case where the velocity of light can be considered as infinitely big, and is contained in the fourth as the limiting case where Planck's constant of action can be considered as infinitely small. The first and partly the third set belong to the fourth as a priori for the description of the experiments. The second set can be connected with any of the other three sets without difficulty and is especially important in its connection with the fourth. The independent existence of the third and fourth sets suggests the existence of a fifth set, of which one, three, and four are limiting cases. This fifth set will probably be found someday in connection with the theory of the elementary particles.

We have omitted from this enumeration the set of concepts connected with the theory of general relativity, since this set has perhaps not yet reached its final form. But it should be emphasized that it is distinctly different from the other four sets.

After this short survey we may come back to the more general question, what one should consider as the characteristic features of such a closed system of axioms and definitions. Perhaps the most important feature is the possibility of finding a consistent mathematical representation for it. This representation must guarantee that the system does not contain contradictions. Then the system must be suited to describe a wide field of experience. The great variety of phenomena in the field should correspond to the great number of solutions of the equations in the mathematical representation. The limitations of the field can generally not be derived from the concepts. The concepts are not sharply defined in their relation to nature, in spite of the sharp definition of their possible connections. The limitations will therefore be found from experience, from the fact that the concepts do not allow a complete description of the observed phenomena.

After this brief analysis of the structure of present-day physics the relation between physics and other branches of natural science may be discussed. The nearest neighbor to physics is chemistry. Actually through quantum theory these two sciences have come to a complete union. But a hundred years ago they were widely separated, their methods of research were quite different, and the concepts of chemistry had at that time no counterpart in physics. Concepts like valency, activity, solubility and volatility had a more qualitative character, and chemistry scarcely belonged to the exact sciences. When the theory of heat had been developed by the middle of the last century scientists started to apply it to the chemical processes, and ever since then the scientific work in this field has been determined by the hope of reducing the laws of chemistry to the mechanics of the atoms. It should be emphasized, however, that this was not possible within the framework of Newtonian mechanics. In order to give a quantitative description of the laws of chemistry one had to formulate a much wider system of concepts for atomic physics. This was finally done in quantum theory, which has its roots just as much in chemistry as in atomic physics. Then it was easy to see that the laws of chemistry could not be reduced to Newtonian mechanics of atomic particles, since the chemical elements displayed in their behavior a degree of stability completely lacking in mechanical systems. But it was not until Bohr's theory of the atom in 1913 that this point had been clearly understood. In the final result, one may say, the concepts of chemistry are in part complementary to the mechanical concepts. If we know that an atom is in its lowest stationary state that determines its chemical properties we cannot at the same time speak about the motion of the electrons in the atom.

The present relation between biology, on the one side, and physics and chemistry, on the other, may be very similar to that between chemistry and physics a hundred years ago. The methods of biology are different from those of physics and chemistry, and the typical biological concepts are of a more qualitative character than those of the exact sciences. Concepts like life, organ, cell, function of an organ, perception have no counterpart in physics or chemistry. On the other hand, most of the progress made in biology during the past hundred years has been achieved through the application of chemistry and physics to the living organism, and the whole tendency of biology in our time is to explain biological phenomena on the basis of the known physical and chemical laws. Again the question arises, whether this hope is justified or not.

Just as in the case of chemistry, one learns from simple biological experience that the living organisms display a degree of stability which general complicated structures consisting of many different types of molecules could certainly not have on the basis of the physical and chemical laws alone. Therefore, something has to be added to the laws of physics and chemistry before the biological phenomena can be completely understood.

With regard to this question two distinctly different views have frequently been discussed in the biological literature. The one view refers to Darwin's theory of evolution in its connection with modern genetics. According to this theory, the only concept which has to be added to those of physics and chemistry in order to understand life is the concept of history. The enormous time interval of roughly four thousand million years that has elapsed since the formation of the earth has given nature the possibility of trying an almost unlimited variety of structures of groups of molecules. Among these structures there have finally been some that could reduplicate themselves by using smaller groups from the surrounding matter, and such structures therefore could be created in great numbers. Accidental changes in the structures provided a still larger variety of the existing structures. Different structures had to compete for the material drawn from the surrounding matter and in this way, through the "survival of the fittest," the evolution of living organisms finally took place. There can be no doubt that this theory contains a very large amount of truth, and many biologists claim that the addition of the concepts of history and evolution to the coherent set of concepts of physics and chemistry will be amply sufficient to account for all biological phenomena. One of the arguments frequently used in favor of this theory emphasizes that wherever the laws of physics and chemistry have been checked in living organisms they have always been found to be correct; there seems definitely to be no place at which some "vital force" different from the forces in physics could enter.

On the other hand, it is just this argument that has lost much of its weight through quantum theory. Since the concepts of physics and chemistry form a closed and coherent set, namely, that of quantum theory, it is necessary that wherever these concepts can be used to describe phenomena the laws connected with the concepts must be valid too. Therefore, wherever one treats living organisms as physicochemical systems, they must necessarily act as such. The only question from which we can learn something about the adequacy of this first view is whether the physicochemical concepts allow a *complete* description of the organisms. Biologists, who answer this question in the negative, generally hold the second view, that has now to be explained.

This second view can perhaps be stated in the following terms: It is very difficult to see how concepts like perception, function of an organ, affection could be a part of the coherent set of the concepts of quantum theory combined with the concept of history. On the other hand, these concepts are necessary for a complete description of life, even if for the moment we exclude mankind as presenting new problems beyond biology. Therefore, it will probably be necessary for an understanding of life to go beyond quantum theory and to construct a new coherent set of concepts, to which physics and chemistry may belong as "limiting cases;" History may be an essential part of it, and concepts like perception, adaptation, affection also will belong to it. If this view is correct, the combination of Darwin's theory with physics and chemistry would not be sufficient to explain organic life; but still it would be true that living organisms can to a large extent be considered as physicochemical systems—as machines, as Descartes and Laplace have put it—and would, if treated as such, also react as such. One could at the same time assume, as Bohr has suggested, that our knowledge of a cell being alive may be complementary to the complete knowledge of its molecular structure. Since a complete knowledge of this structure could possibly be achieved only by operations that destroy the life of the cell, it is logically possible that life precludes the complete determination of its underlying physicochemical structure. Even if one holds this second view one would probably recommend for biological research no other method than has been pursued in the past decades: attempting to explain as much as possible on the basis of the known physicochemical laws, and describing the behavior of organisms carefully and without theoretical prejudices.

The first of these two views is more common among modern biologists than the second; but the experience available at present is certainly not sufficient to decide between the two views. The preference that is given by many biologists to the first view may be due again to the Cartesian partition, which has penetrated so deeply into the human mind during the past centuries. Since the "res cogitans" was confined to men, to the "I," the animals could have no soul, they belonged exclusively to the "res extensa." Therefore, the animals can be understood, so it is argued, on the same terms as matter in general, and the laws of physics and chemistry together with the concept of history must be sufficient to explain their behavior. It is only when the "res cogitans" is brought in that a new situation arises which will require entirely new concepts. But the Cartesian partition is a dangerous oversimplification and it is therefore quite possible that the second view is the correct one.

Quite apart from this question, which cannot be settled yet, we are obviously still very far from such a coherent and closed set of concepts for the description of biological phenomena. The degree of complication in biology is so discouraging that one can at present not imagine any set of concepts in which the connections could be so sharply defined that a mathematical representation could become possible.

If we go beyond biology and include psychology in the discussion, then there can scarcely be any doubt but that the concepts of physics, chemistry, and evolution together will not be sufficient to describe the facts. On this point the existence of quantum theory has changed our attitude from what was believed in the nineteenth century. During that period some scientists were inclined to think that the psychological phenomena could ultimately be explained on the basis of physics and chemistry of the brain. From the quantum-theoretical point of view there is no reason for such an assumption. We would, in spite of the fact that the physical events in the brain belong to the psychic phenomena, not expect that these could be sufficient to explain them. We would never doubt that the brain acts as a physicochemical mechanism if treated as such; but for an understanding of psychic phenomena we would start from the fact that the human mind enters as object and subject into the scientific process of psychology.

Looking back to the different sets of concepts that have been formed in the past or may possibly be formed in the future in the attempt to find our way through the world by means of science, we see that they appear to be ordered by the increasing part played by the subjective element in the set. Classical physics can be considered as that idealization in which we speak about the world as entirely separated from ourselves. The first three sets correspond to this idealization. Only the first set complies entirely with the "a priori" in the philosophy of Kant. In the fourth set, that of quantum theory, man as the subject of science is brought in through the questions which are put to nature in the a priori terms of human science. Quantum theory does not allow a completely objective description of nature. In biology it may be important for a complete understanding that the questions are asked by the species man which itself belongs to the genus of living organisms, in other words, that we already know what life is even before we have defined it scientifically. But one should perhaps not enter into speculations about the possible structure of sets of concepts that have not yet been formed.

When one compares this order with older classifications that belong to earlier stages of natural science one sees that one has now divided the world not into different groups of objects but into different groups of connections. In an earlier period of science one distinquished, for instance, as different groups minerals, plants, animals, men. These objects were taken according to their group as of different natures, made of different materials, and determined in their behavior by different forces. Now we know that it is always the same matter, the same various chemical compounds that may belong to any object, to minerals as well as animals or plants; also the forces that act between the different parts of matter are ultimately the same in every kind of object. What can be distinguished is the kind of connection which is primarily important in a certain phenomenon. For instance, when we speak about the action of chemical forces we mean a kind of connection which is more complicated or in any case different from that expressed in Newtonian mechanics. The world thus appears as a complicated tissue of events, in which connections of different kinds alternate or overlap or combine and thereby determine the texture of the whole.

When we represent a group of connections by a closed and coherent set of concepts, axioms, definitions and laws which in turn is represented by a mathematical scheme we have in fact isolated and idealized this group of connections with the purpose of clarification. But even if complete clarity has been achieved in this way, it is not known how accurately the set of concepts describes reality.

These idealizations may be called a part of the human language that has been formed from the interplay between the world and ourselves, a human response to the challenge of nature. In this respect they may be compared to the different styles of art, say of architecture or music. A style of art can also be defined by a set of formal rules which are applied to the material of this special art. These rules can perhaps not be represented in a strict sense by a set of mathematical concepts and equations, but their fundamental elements are very closely related to the essential elements of mathematics. Equality and inequality, repetition and symmetry, certain group structures play the fundamental role both in art and in mathematics. Usually the work of several generations is needed to develop that formal system which later is called the style of the art, from its simple beginning to the wealth of elaborate forms which characterize its completion. The interest of the artist is concentrated on this process of crystallization, where the material of the art takes, through his action, the various forms that are initiated by the first formal concepts of this style. After the completion the interest must fade again, because the word "interest" means: to be with something, to take part in a process of life, but this process has then come to an end. Here again the question of how far the formal rules of the style represent that reality of life which is meant by the art cannot be decided from the formal rules. Art is always an idealization; the ideal is different from reality—at least from the reality of the shadows, as Plato would have put it—but idealization is necessary for understanding.

This comparison between the different sets of concepts in

natural science with different styles of art may seem very far from the truth to those who consider the different styles of art as rather arbitrary products of the human mind. They would argue that in natural science these different sets of concepts represent objective reality, have been taught to us by nature, are therefore by no means arbitrary, and are a necessary consequence of our gradually increasing experimental knowledge of nature. About these points most scientists would agree; but are the different styles of art an arbitrary product of the human mind? Here again we must not be misled by the Cartesian partition. The style arises out of the interplay between the world and ourselves, or more specifically between the spirit of the time and the artist. The spirit of a time is probably a fact as objective as any fact in natural science, and this spirit brings out certain features of the world which are even independent of time, are in this sense eternal. The artist tries by his work to make these features understandable, and in this attempt he is led to the forms of the style in which he works.

Therefore, the two processes, that of science and that of art, are not very different. Both science and art form in the course of the centuries a human language by which we can speak about the more remote parts of reality, and the coherent sets of concepts as well as the different styles of art are different words or groups of words in this language.

VII.

The Theory of Relativity

WITHIN the field of modern physics the theory of relativity has always played a very important role. It was in this theory that the necessity for a change in the fundamental principles of physics was recognized for the first time. Therefore, a discussion of those problems that had been raised and partly solved by the theory of relativity belongs essentially to our treatment of the philosophical implications of modern physics. In some sense it may be said that —contrary to quantum theory—the development of the theory of relativity from the final recognition of the difficulties to their solution has taken only a very short time. The repetition of Michelson's experiment by Morley and Miller in 1904 was the first definite evidence for the impossibility of detecting the translational motion of the earth by optical methods, and Einstein's decisive paper appeared less than two years later. On the other hand, the experiment of Morley and Miller and Einstein's paper were only the final steps in a development which had started very much earlier and which may be summarized under the heading "electrodynamics of moving bodies."

Obviously the electrodynamics of moving bodies had been an important field of physics and engineering ever since electromotors had been constructed. A serious difficulty had been brought into this subject, however, by Maxwell's discovery of the electromagnetic nature of light waves. These waves differ in one essential property from other waves, for instance, from sound waves: they can be propagated in what seems to be empty space. When a bell rings in a vessel that has been evacuated, the sound does not penetrate to the outside. But light does penetrate easily through the evacuated volume. Therefore, one assumed that light waves could be considered as elastic waves of a very light substance called ether which could be neither seen nor felt but which filled the evacuated space as well as the space in which other matter, like air or glass, existed. The idea that electromagnetic waves could be a reality in themselves, independent of any bodies, did at that time not occur to the physicists. Since this hypothetical substance ether seemed to penetrate through other matter, the question arose: What happens if the matter is set into motion? Does the ether participate in this motion and—if this is the case—how is a light wave propagated in the moving ether?

Experiments which are relevant to this question are difficult for the following reason: The velocities of moving bodies are usually very small compared to the velocity of light. Therefore, the motion of these bodies can only produce very small effects which are proportional to the ratio of the velocity of the body to the velocity of light, or to a higher power of this ratio. Several experiments by Wilson, Rowland, Roentgen and Eichenwald and Fizeau permitted the measurement of such effects with an accuracy corresponding to the first power of this ratio. The theory of the electrons developed by Lorentz in 1895 was able to describe these effects quite satisfactorily. But then the experiment of Michelson, Morley and Miller created a new situation.

This experiment shall be discussed in some detail. In order to get bigger effects and thereby more accurate results, it seemed best to do experiments with bodies of very high velocity. The earth moves around the sun with a velocity of roughly 20 miles/sec. If the ether is at rest with respect to the sun and does not move with the earth, then this fast motion of the ether with respect to the earth should make itself felt in a change of the velocity of light. This velocity should be different depending on whether the light is propagated in a direction parallel or perpendicular to the direction of the motion of the ether. Even if the ether should partly move with the earth, there should be some effect due to what one may call wind of the ether, and this effect would then probably depend on the altitude above sea level at which the experiment is carried out. A calculation of the expected effect showed that it should be very small, since it is proportional to the square of the ratio of the velocity of the earth to that of the light, and that one therefore had to carry out very careful experiments on the interference of two beams of light traveling parallel or perpendicular to the motion of the earth. The first experiment of this kind, carried out by Michelson in 1881, had not been sufficiently accurate. But even later repetitions of the experiment did not reveal the slightest signs of the expected effect. Especially the experiments of Morley and Miller in 1904 could be considered as definite proof that an effect of the expected order of magnitude did not exist.

This result, strange as it was, met another point that had been discussed by the physicists some time before. In Newton's mechanics a certain "principle of relativity" is fulfilled that can be described in the following terms: If in a certain system of reference the mechanical motion of bodies fulfills the laws of Newtonian mechanics, then this is also true for any other frame of reference which is in uniform nonrotating motion with respect to the first system. Or, in other words, a uniform translational motion of a system does not produce any mechanical effects at all and can therefore not be observed by such effects.

Such a principle of relativity—so it seemed to the physicists—could not be true in optics or electrodynamics. If the first system is at rest with respect to the ether, the other systems are not, and therefore their motion with respect to the ether should be recognized by effects of the type considered by Michelson. The negative result of the experiment of Morley and Miller in 1904 revived the idea that such a principle of relativity could be true in electrodynamics as well as Newtonian mechanics.

On the other hand, there was an old experiment by Fizeau in 1851 that seemed definitely to contradict the principle of relativity. Fizeau had measured the velocity of light in a moving liquid. If the principle of relativity was correct, the total velocity of light in the moving liquid should be the sum of the velocity of the liquid and the velocity of light in the liquid at rest. But this was not the case; the experiment of Fizeau showed that the total velocity was somewhat smaller.

Still the negative results of all more recent experiments to recognize the motion "with respect to the ether" inspired the theoretical physicists and mathematicians at that time to look for mathematical interpretations that reconciled the wave equation for the propagation of light with the principle of relativity. Lorentz suggested, in 1904, a mathematical transformation that fulfilled these requirements. He had to introduce the hypothesis that moving bodies are contracted in the direction of motion by a factor depending on the velocity of the body, and in different schemes of reference there are different "apparent" times which in many ways take the place of the "real" time. In this way he could represent something resembling the principle of relativity:

the "apparent" velocity of light was the same in every system of reference. Similar ideas had been discussed by Poincaré, Fitzgerald and other physicists.

The decisive step, however, was taken in the paper by Einstein in 1905 in which he established the "apparent" time of the Lorentz transformation as the "real" time and abolished what had been called "real" time by Lorentz. This was a change in the very foundations of physics; an unexpected and very radical change that required all the courage of a young and revolutionary genius. To take this step one needed, in the mathematical representation of nature, nothing more than the consistent application of the Lorentz transformation. But by its new interpretation the structure of space and time had changed and many problems of physics appeared in a new light. The substance ether, for instance, could be abolished too. Since all systems of reference that are in uniform translation motion with respect to each other are equivalent for the description of nature, there is no meaning in the statement that there is a substance, the ether, which is at rest in only one of these systems. Such a substance is in fact not needed and it is much simpler to say that light waves are propagated through empty space and that electromagnetic fields are a reality of their own and can exist in empty space.

But the decisive change was in the structure of space and time. It is very difficult to describe this change in the words of common language, without the use of mathematics, since the common words "space" and "time" refer to a structure of space and time that is actually an idealization and oversimplification of the real structure. But still we have to try to describe the new structure and we can perhaps do it in the following way:

When we use the term "past" we comprise all those events

which we could know at least in principle, about which we could have heard at least in principle. In a similar manner we comprise by the term "future" all those events which we could influence at least in principle, which we could try to change or to prevent at least in principle. It is not easy for a nonphysicist to see why this definition of the terms "past" and "future" should be the most convenient one. But one can easily see that it corresponds very accurately to our common use of the terms. If we use the terms in this way, it turns out as a result of many experiments that the content of "future" or "past" does not depend on the state of motion or other properties of the observer. We may say that the definition is invariant against the motion of the observer. This is true both in Newtonian mechanics and in Einstein's theory of relativity.

But the difference is this: In classical theory we assume that future and past are separated by an infinitely short time interval which we may call the present moment. In the theory of relativity we have learned that the situation is different: future and past are separated by a finite time interval the length of which depends on the distance from the observer. Any action can only be propagated by a velocity smaller than or equal to the velocity of light. Therefore, an observer can at a given instant neither know of nor influence any event at a distant point which takes place between two characteristic times. The one time is the instant at which a light signal has to be given from the point of the event in order to reach the observer at the instant of observation. The other time is the instant at which a light signal, given by the observer at the instant of the observation, reaches the point of the event. The whole finite time interval between these two instants may be said to belong to the "present time" for the observer at the instant of observation. Any event taking place between the two characteristic times may be called "simultaneous" with the act of observation.

The use of the phrase "may be called" points up an ambiguity in the word "simultaneous," which is due to the fact that this term has been formed from the experience of daily life, in which the velocity of light can always be considered as infinitely high. Actually this term in physics can be defined also in a slightly different manner and Einstein has in his papers used this second definition. When two events happen at the same point in space simultaneously, we say that they coincide; this term is quite unambiguous. Let us now imagine three points in space that lie on a straight line so that the point in the middle has the same distance from each of the two outer points. If two events happen at the two outer points at such times that light signals starting from the events coincide when they reach the point in the middle, we can define the two events as simultaneous. This definition is narrower than the first one. One of its most important consequences is that when two events are simultaneous for one observer they may not be simultaneous for another observer, if he is in motion relative to the first observer. The connection between the two definitions can be established by the statement that whenever two events are simultaneous in the first sense of the term, one can always find a frame of reference in which they are simultaneous in the second sense too.

The first definition of the term "simultaneous" seems to correspond more nearly to its use in daily life, since the question whether two events are simultaneous does in daily life not depend on the frame of reference. But in both relativistic definitions the term has acquired a precision which is lacking in the language of daily life. In quantum theory the physicists had to learn rather early that the terms of classical physics describe nature only inaccurately, that their application is limited by the quantum laws and that one therefore should be cautious in their use. In the theory of relativity the physicists have tried to change the meaning of the words of classical physics, to make the terms more precise in such a way that they fit the new situation in nature.

The structure of space and time that has been brought to light by the theory of relativity has many consequences in different parts of physics. The electrodynamics of moving bodies can be derived at once from the principle of relativity. This principle itself can be formulated as a quite general law of nature pertaining not only to electrodynamics or mechanics but to any group of laws: The laws take the same form in all systems of reference, which are different from each other only by a uniform translational motion; they are invariant against the Lorentz transformation.

Perhaps the most important consequence of the principle of relativity is the inertia of energy, or the equivalence of mass and energy. Since the velocity of light is the limiting velocity which can never be reached by any material body, it is easy to see that it is more difficult to accelerate a body that is already moving very fast than a body at rest. The inertia has increased with the kinetic energy. But quite generally any kind of energy will, according to the theory of relativity, contribute to the inertia, i.e., to the mass, and the mass belonging to a given amount of energy is just this energy divided by the square of the velocity of light. Therefore, every energy carries mass with it; but even a rather big energy carries only a very small mass, and this is the reason why the connection between mass and energy had not been observed before. The two laws of the conservation of mass and the conservation of charge lose their separate validity and are combined into one single law which may be called the law of conservation of energy or mass. Fifty years ago, when the theory of relativity was formulated, this hypothesis of the equivalence of mass and energy seemed to be a complete revolution in physics, and there was still very little experimental evidence for it. In our times we see in many experiments how elementary particles can be created from kinetic energy, and how such particles are annihilated to form radiation; therefore, the transmutation from energy into mass and vice versa suggests nothing unusual. The enormous release of energy in an atomic explosion is another and still more spectacular proof of the correctness of Einstein's equation. But we may add here a critical historical remark.

It has sometimes been stated that the enormous energies of atomic explosions are due to a direct transmutation of mass into energy, and that it is only on the basis of the theory of relativity that one has been able to predict these energies. This is, however, a misunderstanding. The huge amount of energy available in the atomic nucleus was known ever since the experiments of Becquerel, Curie and Rutherford on radioactive decay. Any decaying body like radium produces an amount of heat about a million times greater than the heat released in a chemical process in a similar amount of material. The source of energy in the fission process of uranium is just the same as that in the α-decay of radium, namely, mainly the electrostatic repulsion of the two parts into which the nucleus is separated. Therefore, the energy of an atomic explosion comes directly from this source and is not derived from a transmutation of mass into energy. The number of elementary particles with finite rest mass does not decrease during the explosion. But it is true that the binding energies of the particles in an atomic nucleus do show up in their masses and therefore the release of energy is in this indirect manner also connected with changes in the masses of the nuclei. The equivalence of mass and energy has—besides its great importance in physics—also raised problems concerning very old philosophical questions. It has been the thesis of several philosophical systems of the past that substance or matter cannot be destroyed. In modern physics, however, many experiments have shown that elementary particles, e.g., positrons and electrons, can be annihilated and transmuted into radiation. Does this mean that these older philosophical systems have been disproved by modern experience and that the arguments brought forward by the earlier systems have been misleading?

This would certainly be a rash and unjustified conclusion, since the terms "substance" and "matter" in ancient or medieval philosophy cannot simply be identified with the term "mass" in modern physics. If one wished to express our modern experience in the language of older philosophies, one could consider mass *and* energy as two different forms of the same "substance" and thereby keep the idea of substance as indestructible.

On the other hand, one can scarcely say that one gains much by expressing modern knowledge in an old language. The philosophic systems of the past were formed from the bulk of knowledge available at their time and from the lines of thought to which such knowledge had led. Certainly one should not expect the philosophers of many hundreds of years ago to have foreseen the development of modern physics or the theory of relativity. Therefore, the concepts to which the philosophers were led in the process of intellectual clarification a long time ago cannot possibly be adapted to phenomena that can only be observed by the elaborate technical tools of our time.

But before going into a discussion of philosophical implications of the theory of relativity its further development has to be described.

The hypothetical substance "ether," which had played such an important role in the early discussions on Maxwell's theories in the nineteenth century, had—as has been said before—been abolished by the theory of relativity. This is sometimes stated by saying that the idea of absolute space has been abandoned. But such a statement has to be accepted with great caution. It is true that one cannot point to a special frame of reference in which the substance ether is at rest and which could therefore deserve the name "absolute space." But it would be wrong to say that space has now lost all of its physical properties. The equations of motion for material bodies or fields still take a different form in a "normal" system of reference from another one which rotates or is in a nonuniform motion with respect to the "normal" one. The existence of centrifugal forces in a rotating system proves—so far as the theory of relativity of 1905 and 1906 is concerned—the existence of physical properties of space which permit the distinction between a rotating and a nonrotating system.

This may not seem satisfactory from a philosophical point of view, from which one would prefer to attach physical properties only to physical entities like material bodies or fields and not to empty space. But so far as the theory of electromagnetic processes or mechanical motions is concerned, this existence of physical properties of empty space is simply a description of facts that cannot be disputed.

A careful analysis of this situation about ten years later, in 1916, led Einstein to a very important extension of the theory of relativity, which is usually called the theory of "general relativity." Before going into a description of the main ideas of this new theory it may be useful to say a few words about the degree of certainty with which we can rely on the correctness of these two parts of the theory of relativity. The theory of 1905 and 1906 is based on a very great number of well-established facts: on the experiments of Michelson and Morley and many similar ones, on the equivalence of mass and energy in innumerable radioactive processes, on the dependence of the lifetime of radioactive bodies on their velocity, etc. Therefore, this theory belongs to the firm foundations of modern physics and cannot be disputed in our present situation.

For the theory of general relativity the experimental evidence is much less convincing, since the experimental material is very scarce. There are only a few astronomical observations which allow a checking of the correctness of the assumptions. Therefore, this whole theory is more hypothetical than the first one.

The cornerstone of the theory of general relativity is the connection between inertia and gravity. Very careful measurements have shown that the mass of a body as a source of gravity is exactly proportional to the mass as a measure for the inertia of the body. Even the most accurate measurements have never shown any deviation from this law. If the law is generally true, the gravitational forces can be put on the same level with the centrifugal forces or with other forces that arise as a reaction of the inertia. Since the centrifugal forces had to be considered as due to physical properties of empty space, as had been discussed before, Einstein turned to the hypothesis that the gravitational forces also are due to properties of empty space. This was a very important step which necessitated at once a second step of equal importance. We know that the forces of gravity are produced by masses. If therefore gravitation is connected with properties of space, these properties of space must be caused or influenced by the masses. The centrifugal forces in a rotating system must be produced by the rotation (relative to the system) of possibly very distant masses.

In order to carry out the program outlined in these few sentences Einstein had to connect the underlying physical ideas with the mathematical scheme of general geometry that had been developed by Riemann. Since the properties of space seemed to change continuously with the gravitational fields, its geometry had to be compared with the geometry on curved surfaces where the straight line of Euclidean geometry has to be replaced by the geodetical line, the line of shortest distance, and where the curvature changes continuously. As a final result Einstein was able to give a mathematical formulation for the connection between the distribution of masses and the determining parameters of the geometry. This theory did represent the common facts about gravitation. It was in a very high approximation identical with the conventional theory of gravitation and predicted furthermore a few interesting effects which were just at the limit of measurability. There was, for instance, the action of gravity on light. When monochromatic light is emitted from a heavy star, the light quanta lose energy when moving away through the gravitational field of the star; a red shift of the emitted spectral line follows. There is as yet no experimental evidence for this red shift, as the discussion of the experiments by Freundlich has clearly shown. But it would also be premature to conclude that the experiments contradict the prediction of Einstein's theory. A beam of light that passes near the sun should be deflected by its gravitational field. The deflection has been found experimentally by Freundlich in the right order of magnitude; but whether the deflection agrees quantitatively with the value predicted by Einstein's theory has not yet been decided. The best evidence for the validity of the theory of general relativity seems to be the procession in the orbital motion of the planet Mercury, which apparently is in very good agreement with the value predicted by the theory.

Though the experimental basis of general relativity is still rather narrow, the theory contains ideas of the greatest importance. During the whole period from the mathematicians of ancient Greece to the nineteenth century, Euclidean geometry had been considered as evident; the axioms of Euclid were regarded as the foundation of any mathematical geometry, a foundation that could not be disputed. Then, in the nineteenth century, the mathematicians Bolyai and Lobachevsky, Gauss and Riemann found that other geometries could be invented which could be developed with the same mathematical precision as that of Euclid; therefore, the questions as to which geometry was correct turned out to be an empirical one. But it was only through the work of Einstein that the question could really be taken up by the physicists. The geometry discussed in the theory of general relativity was not concerned with three-dimensional space only but with the four-dimensional manifold consisting of space and time. The theory established a connection between the geometry in this manifold and the distribution of masses in the world. Therefore, this theory raised in an entirely new form the old questions of the behavior of space and time in the largest dimensions; it could suggest possible answers that could be checked by observations.

Consequently, very old philosophic problems were taken up that had occupied the mind of man since the earliest phases of philosophy and science. Is space finite or infinite? What was there before the beginning of time? What will happen at the end of time? Or is there no beginning and no end? These questions had found different answers in different philosophies and religions. In the philosophy of Aristotle, for instance, the total space of the universe was finite (though it was infinitely divisible). Space was due to the extension of bodies, it was connected with the bodies; there was no space where there were no bodies. The universe consisted of the earth and the sun and the stars: a finite number of bodies. Beyond the sphere of the stars there was no space; therefore, the space of the universe was finite.

In the philosophy of Kant this question belonged to what he called "antinomies"—questions that cannot be answered, since two different arguments lead to opposite results. Space cannot be finite, since we cannot imagine that there should be an end to space; to whichever point in space we come we can always imagine that we can go beyond. At the same time space cannot be infinite, because space is something that we can imagine (else the word "space" would not have been formed) and we cannot imagine an infinite space. For this second thesis the argument of Kant has not been verbally reproduced. The sentence "space is infinite" means for us something negative; we cannot come to an end of space. For Kant it means that the infinity of space is really given, that it "exists" in a sense that we can scarcely reproduce. Kant's result is that a rational answer to the question whether space is finite or infinite cannot be given because the whole universe cannot be the object of our experience.

A similar situation is found with respect to the problem of the infinity of time. In the *Confessions* of St. Augustine, for instance, this question takes the form: What was God doing before He created the world? Augustine is not satisfied with the joke: "God was busy preparing Hell for those who ask foolish questions." This, he says, would be too cheap an answer, and he tries to give a rational analysis of the problem. Only for us is time passing by; it is expected by us as future; it passes by as the present moment and is remembered by us as past. But God is not in time; a thousand years are for Him as one day, and one day as a thousand years. Time has been created together with the world, it belongs to the world, therefore time did not exist before the universe existed. For God the whole course of the universe is given at once. There was no time before He created the world. It is obvious that in such statements the word "created" at once raises all the essential difficulties. This word as it is usually understood means that something has come into being that has not been before, and in this sense it presupposes the concept of time. Therefore, it is impossible to define in rational terms what could be meant by the phrase "time has been created." This fact reminds us again of the often discussed lesson that has been learned from modern physics: that every word or concept, clear as it may seem to be, has only a limited range of applicability.

In the theory of general relativity these questions about the infinity of space and time can be asked and partly answered on an empirical basis. If the connection between the four-dimensional geometry in space and time and the distribution of masses in the universe has been correctly given by the theory, then the astronomical observations on the distribution of galaxies in space give us information about the geometry of the universe as a whole. At least one can build "models" of the universe, cosmological pictures, the consequences of which can be compared with the empirical facts.

From the present astronomical knowledge one cannot definitely distinguish between several possible models. It may be that the space filled by the universe is finite. This would not mean that there is an end of the universe at some place. It would only mean that by proceeding farther and farther in one direction in the universe one would finally come back to the point from which one had started. The situation would be similar as in the two-dimensional geometry on the surface of the earth where we, when starting from a point in an eastward direction, finally come back to this point from the west.

With respect to time there seems to be something like a beginning. Many observations point to an origin of the universe about four billion years ago; at least they seem to show that at that time all matter of the universe was concentrated in a much smaller space than it is now and has expanded ever since from this small space with different velocities. The same time of four billion years is found in many different observations (e.g., from the age of meteorites, of minerals on the earth, etc.), and therefore it would be difficult to find an interpretation essentially different from this idea of an origin. If it is the correct one it would mean that beyond this time the concept of time would undergo essential changes. In the present state of astronomical observations the questions about the space-time geometry on a large scale cannot yet be answered with any degree of certainty. But it is extremely interesting to see that these questions may possibly be answered eventually on a solid empirical basis. For the time being even the theory of general relativity rests on a very narrow experimental foundation and must be considered as much less certain than the so-called theory of special relativity expressed by the Lorentz transformation.

Even if one limits the further discussions of this latter theory there is no doubt that the theory of relativity has deeply changed our views on the structure of space and time. The most exciting aspect of these changes is perhaps not their special nature but the fact that they have been possible. The structure of space and time which had been defined by Newton as the basis of his mathematical description of nature was simple and consistent and corresponded very closely to the use of the concepts space and time in daily life. This correspondence was in fact so close that Newton's definitions could be considered as the precise mathematical formulation of these common concepts. Before the theory of relativity it seemed completely obvious that events could be ordered in time independent of their location in space. We know now that this impression is created in daily life by the fact that the velocity of light is so very much higher than any other velocity occurring in practical experience; but this restriction was of course not realized at that time. And even if we know the restriction now we can scarcely imagine that the time order of events should depend on their location.

The philosophy of Kant later on drew attention to the fact that the concepts of space and time belong to our relation to nature, not to nature itself; that we could not describe nature without using these concepts. Consequently, these concepts are "a priori' in some sense, they are the condition for and not primarily the result of experience, and it was generally believed that they could not be touched by new experience. Therefore, the necessity of the change appeared as a great surprise. It was the first time that the scientists learned how cautious they had to be in applying the concepts of daily life to the refined experience of modern experimental science. Even the precise and consistent formulation of these concepts in the mathematical language of Newton's mechanics or their careful analysis in the philosophy of Kant had offered no protection against the critical analysis possible through extremely accurate measurements. This warning later proved extremely useful in the development of modern physics, and it would certainly have been still more difficult to understand quantum theory had not the success of the theory of relativity warned the physicists against the uncritical use of concepts taken from daily life or from classical physics.

VIII.

Criticism and Counterproposals to the Copenhagen Interpretation of Quantum Theory

THE Copenhagen interpretation of quantum theory has led the physicists far away from the simple materialistic views that prevailed in the natural science of the nineteenth century. Since these views had not only been intrinsically connected with natural science of that period but had also found a systematic analysis in some philosophic systems and had penetrated deeply into the mind even of the common men on the street, it can be well understood that many attempts have been made to criticize the Copenhagen interpretation and to replace it by one more in line with the concepts of classical physics or materialistic philosophy.

These attempts can be divided into three different groups. The first group does not want to change the Copenhagen interpretation so far as predictions of experimental results are concerned; but it tries to change the language of this interpretation in order to get a closer resemblance to classical physics. In other words, it tries to change the philosophy without changing the physics. Some papers of this first group restrict their agreement with the experimental predictions of the Copenhagen interpretation to all those experiments that have hitherto been carried out or that belong to normal electronic physics.

The second group realizes that the Copenhagen interpretation is the only adequate one, if the experimental results agree everywhere with the predictions of this interpretation. Therefore, the papers of this group try to change quantum theory to some extent in certain critical points.

The third group, finally, expresses rather its general dissatisfaction with the results of the Copenhagen interpretation and especially with its philosophical conclusions, without making definite counterproposals. Papers by Einstein, von Laue and Schrödinger belong to this third group which has historically been the first of the three groups.

However, all the opponents of the Copenhagen interpretation do agree on one point. It would, in their view, be desirable to return to the reality concept of classical physics or, to use a more general philosophic term, to the ontology of materialism. They would prefer to come back to the idea of an objective real world whose smallest parts exist objectively in the same sense as stones or trees exist, independently of whether or not we observe them.

This, however, is impossible or at least not entirely possible because of the nature of the atomic phenomena, as has been discussed in some of the earlier chapters. It cannot be our task to formulate wishes as to how the atomic phenomena should be; our task can only be to understand them.

When one analyzes the papers of the first group, it is important to realize from the beginning that their interpretations cannot be refuted by experiment, since they only repeat the Copenhagen interpretation in a different language. From a strictly positivistic standpoint one may even say that we are here concerned not with counterproposals to the Copenhagen interpretation but with its exact repetition in a different language. Therefore, one can only dispute the suitability of this language. One group of counterproposals works with the idea of "hidden parameters." Since the quantum-theoretical laws determine in general the results of an experiment only statistically, one would from the classical standpoint be inclined to think that there exist some hidden parameters which escape observation in any ordinary experiment but which determine the outcome of the experiment in the normal causal way. Therefore, some papers try to construct such parameters within the framework of quantum mechanics.

Along this line, for instance, Bohm has made a counterproposal to the Copenhagen interpretation, which has recently been taken up to some extent also by de Broglie. Bohm's interpretation has been worked out in detail. It may therefore serve here as a basis for the discussions. Bohm considers the particles as "objectively real" structures, like the point masses in Newtonian mechanics. The waves in configuration space are in his interpretation "objectively real" too, like electric fields. Configuration space is a space of many dimensions referring to the different co-ordinates of all the particles belonging to the system. Here we meet a first difficulty: what does it mean to call waves in configuration space "real"? This space is a very abstract space. The word "real" goes back to the Latin word "res," which means "thing"; but things are in the ordinary three-dimensional space, not in an abstract configuration space. One may call the waves in configuration space "objective" when one wants to say that these waves do not depend on any observer; but one can scarcely call them "real" unless one is willing to change the meaning of the word. Bohm goes on defining the lines perpendicular to the surfaces of constant wave-phase as the possible orbits of the particles. Which of these lines is the "real" orbit depends, according to him, on the history of the system and the measuring apparatus and cannot be decided without knowing more about the system and the measuring equipment than actually can be known. This history contains in fact the hidden parameters, the "actual orbit" before the experiment started.

One consequence of this interpretation is, as Pauli has emphasized, that the electrons in the ground states of many atoms should be at rest, not performing any orbital motion around the atomic nucleus. This looks like a contradiction of the experiments, since measurements of the velocity of the electrons in the ground state (for instance, by means of the Compton effect) reveal always a velocity distribution in the ground state, which is—in conformity with the rules of quantum mechanics—given by the square of the wave function in momentum or velocity space. But here Bohm can argue that the measurement can no longer be evaluated by the ordinary laws. He agrees that the normal evaluation of the measurement would indeed lead to a velocity distribution; but when the quantum theory for the measuring equipment is taken into account—especially some strange quantum potentials introduced ad hoc by Bohm—then the statement is admissible that the electrons "really" always are at rest. In measurements of the position of the particle, Bohm takes the ordinary interpretation of the experiments as correct; in measurements of the velocity he rejects it. At this price Bohm considers himself able to assert: "We do not need to abandon the precise, rational and objective description of individual systems in the realm of quantum theory." This objective description, however, reveals itself as a kind of "ideological superstructure," which has little to do with immediate physical reality; for the hidden parameters of Bohm's interpretation are of such a kind that they can never occur in the description of real processes, if quantum theory remains unchanged.

In order to escape this difficulty, Bohm does in fact express the hope that in future experiments in the range of the elementary particles the hidden parameters may yet play a physical part, and that quantum theory may thus be proved false. When such strange hopes were expressed, Bohr used to say that they were similar in structure to the sentence: "We may hope that it will later turn out that sometimes $2 \times 2 = 5$, for this would be of great advantage for our finances." Actually the fulfillment of Bohm's hopes would cut the ground from beneath not only quantum theory but also Bohm's interpretation. Of course it must at the same time be emphasized that the analogy just mentioned, although complete, does not represent a logically compelling argument against a possible future alteration of quantum theory in the manner suggested by Bohm. For it would not be fundamentally unimaginable that, for example, a future extension of mathematical logic might give a certain meaning to the statement that in exceptional cases $2 \times 2 = 5$, and it might even be possible that this extended mathematics would be of use in calculations in the field of economics. We are nevertheless actually convinced, even without cogent logical grounds, that such changes in mathematics would be of no help to us financially. Therefore, it is very difficult to understand how the mathematical proposals which the work of Bohm indicates as a possible realization of his hopes could be used for the description of physical phenomena.

If we disregard this possible alteration of quantum theory, then Bohm's language, as we have already pointed out, says nothing about physics that is different from what the Copenhagen interpretation says. There then remains only the question of the suitability of this language. Besides the objection already made that in speaking of particle orbits we are concerned with a superfluous "ideological superstructure," it must be particularly mentioned here that Bohm's language destroys the symmetry between position and velocity which is implicit in quantum theory; for the measurements of position Bohm accepts the usual interpretation, for the measurements of velocity or momentum he rejects it. Since the symmetry properties always constitute the most essential features of a theory, it is difficult to see what would be gained by omitting them in the corresponding language. Therefore, one cannot consider Bohm's counterproposal to the Copenhagen interpretation as an improvement.

A similar objection can be raised in a somewhat different form against the statistical interpretations put forward by Bopp and (on a slightly different line) by Fenyes. Bopp considers the creation or the annihilation of a particle as the fundamental process of quantum theory, the particle is "real" in the classical sense of the word, in the sense of materialistic ontology, and the laws of quantum theory are considered as a special case of correlation statistics for such events of creation and annihilation. This interpretation, which contains many interesting comments on the mathematical laws of quantum theory, can be carried out in such a manner that it leads, as regards the physical consequences, to exactly the same conclusions as the Copenhagen interpretation. So far it is, in the positivistic sense, isomorphic with it, as is Bohm's. But in its language it destroys the symmetry between particles and waves that otherwise is a characteristic feature of the mathematical scheme of quantum theory. As early as 1928 it was shown by Jordan, Klein and Wigner that the mathematical scheme can be interpreted not only as a quantization of particle motion but also as a quantization of three-dimensional matter waves; therefore, there is no reason to consider these matter waves as less real than the particles. The symmetry between waves and particles could be ensured in Bopp's interpretation only if the corresponding correlation statistics were developed for matter waves in space and time as well, and if the question was left open whether particles or waves are to be considered as the "actual" reality.

The assumption that particles are real in the sense of the materialistic ontology will always lead to the temptation to consider deviations from the uncertainty principle as "basically" posscible. Fenyes, for instance, says that "the existence of the uncertainty principle [which he connects with certain statistical relations] by no means renders impossible the simultaneous measurement, with arbitrary accuracy, of position and velocity." Fenyes does not, however, state how such measurements should be carried out in practice, and therefore his considerations seem to remain abstract mathematics.

Weizel, whose counterproposals to the Copenhagen interpretation are akin to those of Bohm and Fenyes, relates the "hidden parameters" to a new kind of particle introduced ad hoc, the "zeron," which is not otherwise observable. However, such a concept runs into the danger that the interaction between the real particles and the zerons dissipates the energy among the many degrees of freedom of the zeron field, so that the whole of thermodynamics becomes a chaos. Weizel has not explained how he hopes to avoid this danger.

The standpoint of the entire group of publications mentioned so far can perhaps best be defined by recalling a similar discussion of the theory of special relativity. Anyone who was dissatisfied with Einstein's negation of the ether, of absolute space and of absolute time could then argue as follows: The non-existence of absolute space and absolute time is by no means proved by the theory of special relativity. It has been shown only that true space and true time do not occur directly in any ordinary experiment; but if this aspect of the laws of nature has been correctly taken into account, and thus the correct "apparent" times have been introduced for moving co-ordinate systems, there would be no arguments against the assumption of an absolute space. It would even be plausible to assume that the center of gravity of our galaxy is (at least approximately) at rest in absolute space. The critic of the special theory of relativity might add that we may hope that future measurements will allow the unambiguous definition of absolute space (that is, of the "hidden parameter" of the theory of relativity) and that the theory of relativity will thus be refuted.

It is seen at once that this argument cannot be refuted by experiment, since it as yet makes no assertions which differ from those of the theory of special relativity. But such an interpretation would destroy in the language used the decisive symmetry property of the theory, namely, the Lorentz invariance, and it must therefore be considered inappropriate.

The analogy to quantum theory is obvious. The laws of quantum theory are such that the "hidden parameters," invented ad hoc, can never be observed. The decisive symmetry properties are thus destroyed if we introduce the hidden parameters as a fictitious entity into the interpretation of the theory.

The work of Blochinzev and Alexandrov is quite different in its statement of the problem from those discussed before. These authors expressly and from the beginning restrict their objections against the Copenhagen interpretation to the philosophical side of the problem. The physics of this interpretation is accepted unreservedly.

The external form of the polemic, however, is so much the sharper: "Among the different idealistic trends in contemporary physics the so-called Copenhagen school is the most reactionary. The present article is devoted to the unmasking of the idealistic and agnostic speculations of this school on the basic problems of quantum physics," writes Blochinzev in his introduction. The acerbity of the polemic shows that here we have to do not with science alone but with a confession of faith, with adherence to a certain creed. The aim is expressed at the end with a quotation from the work of Lenin: "However marvellous, from the point of view of the common human intellect, the transformation of the unweighable ether into weighable material, however strange the electrons lack of any but electromagnetic mass, however unusual the restriction of the mechanical laws of motion to but one realm of natural phenomena and their subordination to the deeper laws of electromagnetic phenomena, and so on—all this is but another *confirmation* of dialectic materialism." This latter statement seems to make Blochinzev's discussion about the relation of quantum theory to the philosophy of dialectic materialism less interesting in so far as it seems to degrade it to a staged trial in which the verdict is known before the trial has begun. Still it is important to get complete clarity about the arguments brought forward by Blochinzev and Alexandrov.

Here, where the task is to rescue materialistic ontology, the attack is chiefly made against the introduction of the observer into the interpretation of quantum theory. Alexandrov writes: "We must therefore understand by 'result of measurement' in quantum theory only the objective effect of the interaction of the electron with a suitable object. Mention of the observer must be avoided, and we must treat objective conditions and objective effects. A physical quantity is an objective characteristic of the phenomenon, but not the result of an observation." According to Alexandrov, the wave function in configuration space characterizes the objective state of the electron.

In his presentation Alexandrov overlooks the fact that the formalism of quantum theory does not allow the same degree of objectivation as that of classical physics. For instance, if the interaction of a system with the measuring apparatus is treated as a whole according to quantum mechanics and if both are regarded as cut off from the rest of the world, then the formalism of quantum theory does not as a rule lead to a definite result; it will not lead, e.g., to the blackening of the photographic plate at a given point. If one tries to rescue Alexandrov's "objective effect" by saying that "in reality" the plate is blackened at a given point after the interaction, the rejoinder is that the quantum mechanical treatment of the closed system consisting of electron, measuring apparatus and plate is no longer being applied. It is the "factual" character of an event describable in terms of the concepts of daily life which is not without further comment contained in the mathematical formalism of quantum theory, and which appears in the Copenhagen interpretation by the introduction of the observer. Of course the introduction of the observer must not be misunderstood to imply that some kind of subjective features are to be brought into the description of nature. The observer has, rather, only the function of registering decisions, i.e., processes in space and time, and it does not matter whether the observer is an apparatus or a human being; but the registration, i.e., the transition from the "possible" to the "actual," is absolutely necessary here and cannot be omitted from the interpretation of quantum theory. At this point quantum theory is intrinsically connected with thermodynamics in so far as every act of observation is by its very nature an irreversible process; it is only through such irreversible processes that the formalism of quantum theory can be consistently connected with actual events in space and time. Again the irreversibility is—when projected into the mathematical representation of the phenomena—a consequence of the observer's incomplete knowledge of the system and in so far not completely "objective."

Blochinzev formulates matter slightly differently from Alexandrov: "In quantum mechanics we describe not a state of the particle in itself but the fact that the particle belongs to this or that statistical assembly. This belonging is completely objective and does not depend on statements made by the observer." Such formulations, however, take us very far—probably too far—away from materialistic ontology. To make this point clear it is useful to recall how this belonging to a statistical assembly is used in the interpretation of classical thermodynamics. If an observer has determined the temperature of a system and wants to draw from his results conclusions about the molecular motions in the system he is able to say that the system is just one sample out of a canonical ensemble and thus he may consider it as possibly having different energies. "In reality"—so we would conclude in classical physics—the system has only one definite energy at a given time, and none of the others is realized. The observer has been deceived if he considered a different energy at that moment as possible. The canonical ensemble contains statements not only about the system itself but also about the observer's incomplete knowledge of the system. If Blochinzev in quantum theory tries to call a system's belonging to an assembly "completely objective," he uses the word "objective" in a different sense from that in classical physics. For in classical physics this belonging means, as has been said, statements not only about the system but also about the observer's degree of knowledge of the system. One exception must be made to this assertion in quantum theory. If in quantum theory the assembly is characterized by only one wave function in configuration space (and not, as usual, by a statistical matrix), we meet a special situation (the so-called "pure case") in which the description can be called objective in some sense and in which the element of incomplete knowledge does not occur immediately. But since every measurement would (on account of its irreversible features) reintroduce the element of incomplete knowledge, the situation would not be fundamentally different.

Above all, we see from these formulations how difficult it is when we try to push new ideas into an old system of concepts belonging to an earlier philosophy—or, to use an old metaphor, when we attempt to put new wine into old bottles. Such attempts are always distressing, for they mislead us into continually occupying ourselves with the inevitable cracks in the old bottles instead of rejoicing over the new wine. We cannot possibly expect those thinkers who a century ago introduced dialectic materialism to have foreseen the development of quantum theory. Their concepts of matter and reality could not possibly be adapted to the results of the refined experimental technique of our days.

Perhaps one should add at this point some general remarks about the attitude of the scientist to a special creed; it may be a religious or a political creed. The fundamental difference between the religious and the political creed—that the latter refers to the immediate material reality of the world around us, while the former has as its object another reality beyond the material world—is not important for this special question; it is the problem of creed itself that is to be discussed. From what has been said one would be inclined to demand that the scientist should never rely on special doctrines, never confine his method of thinking to a special philosophy. He should always be prepared to have the foundations of his knowledge changed by new experience. But this demand would again be an oversimplification of our situation in life for two reasons. The first is that the structure of our thinking is determined in our youth by ideas which we meet at that time or by getting into contact with strong personalities from whom we learn. This structure will form an integrating part of all our later work and it may well make it difficult for us to adapt ourselves to entirely different ideas later on. The second reason is that we belong to a community or a society. This community is kept together by common ideas, by a common scale of ethical values, or by a common language in which one speaks about the general problems of life. The common ideas may be supported by the authority of a church, a party or the state and, even if this is not the case, it may be difficult to go away from the common ideas without getting into conflict with the community. Yet the results of scientific thinking may contradict some of the common ideas. Certainly it would be unwise to demand that the scientist should generally not be a loyal member of his community, that he should be deprived of the happiness that may come from belonging to a community, and it would be equally unwise to desire that the common ideas of society which from the scientific point of view are always simplifications should change instantaneously with the progress of scientific knowledge, that they should be as variable as scientific theories must necessarily be. Therefore, at this point we come back even in our time to the old problem of the "twofold truth" that has filled the history of Christian religion throughout the later Middle Ages. There is the very disputable doctrine that "positive religion—whatever form it may take—is an indispensable need for the mass of the people, while the man of science seeks the real truth back of religion and seeks it only there." "Science is esoteric," so it is said, "it is only for the few." If in our time political doctrines and social activities take the part of positive religion in some countries, the problem is still essentially the same. The scientist's first claim will always be intellectual honesty, while the community will frequently ask of the scientist that—in view of the variability of science—he at least wait a few decades before expressing in public his dissenting opinions. There is probably no simple solution to this problem, if tolerance alone is not sufficient; but some consolation may come from the fact that it is certainly an old problem belonging to human life.

Coming back now to the counterproposals to the Copenhagen interpretation of quantum theory we have to discuss the second group of proposals, which try to change quantum theory in order to arrive at a different philosophical interpretation. The most careful attempt in this direction has been made by Janossy, who has realized that the rigorous validity of quantum mechanics compels us to depart from the reality concept of classical physics. He therefore seeks to alter quantum mechanics in such a way that, although many of the results remain true, its structure approaches that of classical physics. His point of attack is what is called "the reduction of wave packets," i.e., the fact that the wave function or, more generally, the probability function changes discontinuously when the observer takes cognizance of a result of measurement. Janossy notices that this reduction cannot be deduced from the differential equations of the mathematical formalism and he believes that he can conclude from this that there is an inconsistency in the usual interpretation. It is well known that the "reduction of wave packets" always appears in the Copenhagen interpretation when the transition is completed from the possible to the actual. The probability function, which covered a wide range of possibilities, is suddenly reduced to a much narrower range by the fact that the experiment has led to a definite result, that actually a certain event has happened. In the formalism this reduction requires that the so-called interference of probabilities, which is the most characteristic phenomenon of quantum theory, is destroyed by the partly undefinable and irreversible interactions of the system with the measuring apparatus and the rest of the world. Janossy now tries to alter quantum mechanics by the introduction of so-called damping terms into the equations, in such a way that the interference terms disappear of themselves after a finite time. Even if this corresponds to reality—and there is no reason to suppose this from the experiments that have been performed—there would still remain a number of alarming consequences of such an interpretation, as Janossy himself points out (e.g., waves which are propagated faster than the velocity of light, interchange of the time sequence of cause and effect, etc.). Therefore, we should hardly be ready to sacrifice the simplicity of quantum theory for this kind of view until we are compelled by experiments to do so.

Among the remaining opponents of what is sometimes called the "orthodox" interpretation of quantum theory, Schrödinger has taken an exceptional position inasmuch as he would ascribe the "objective reality" not to the particles but to the waves and is not prepared to interpret the waves as "probability waves only." In his paper entitled "Are There Quantum Jumps?" he attempts to deny the existence of quantum jumps altogether (one may question the suitability of the term "quantum jump" at this place and could replace it by the less provocative term "discontinuity"). Now, Schrödinger's work first of all contains some misunderstanding of the usual interpretation. He overlooks the fact that only the waves in configuration space (or the "transformation matrices") are probability waves in the usual interpretation, while the three-dimensional matter waves or radiation waves are not. The latter have just as much and just as little "reality" as the particles; they have no direct connection with probability waves but have a continuous density of energy and momentum, like an electromagnetic field in Maxwell's theory. Schrödinger therefore rightly emphasizes that at this point the processes can be conceived of as being more continuous than they usually are. But this interpretation cannot remove the element of discontinuity that is found everywhere in atomic physics; any scintillation screen or Geiger counter demonstrates this element at once. In the usual interpretation of quantum theory it is contained in the transition from the possible to the actual. Schrödinger himself makes no counterproposal as to how he intends to introduce the element of discontinuity, everywhere observable, in a different manner from the usual interpretation.

Finally, the criticism which Einstein, Laue and others have expressed in several papers concentrates on the question whether the Copenhagen interpretation permits a unique, objective description of the physical facts. Their essential arguments may be stated in the following terms: The mathematical scheme of quantum theory seems to be a perfectly adequate description of the statistics of atomic phenomena. But even if its statements about the probability of atomic events are completely correct, this interpretation does not describe what actually happens independently of or between the observations. But something must happen, this we cannot doubt; this something need not be described in terms of electrons or waves or light quanta, but unless it is described somehow the task of physics is not completed. It cannot be admitted that it refers to the act of observation only. The physicist must postulate in his science that he is studying a world which he himself has not made and which would be present, essentially unchanged, if he were not there. Therefore, the Copenhagen interpretation offers no real understanding of the atomic phenomena.

It is easily seen that what this criticism demands is again the old materialistic ontology. But what can be the answer from the point of view of the Copenhagen interpretation?

We can say that physics is a part of science and as such aims at a description and understanding of nature. Any kind of understanding, scientific or not, depends on our language, on the communication of ideas. Every description of phenomena, of experiments and their results, rests upon language as the only means of communication. The words of this language represent the concepts of daily life, which in the scientific language of physics may be refined to the concepts of classical physics. These concepts are the only tools for an unambiguous communication about events, about the setting up of experiments and about their results. If therefore the atomic physicist is asked to give a description of what really happens in his experiments, the words "description" and "really" and "happens" can only refer to the concepts of daily life or of classical physics. As soon as the physicist gave up this basis he would lose the means of unambiguous communication and could not continue in his science. Therefore, any statement about what has "actually happened" is a statement in terms of the classical concepts and —because of thermodynamics and of the uncertainty relations— by its very nature incomplete with respect to the details of the atomic events involved. The demand to "describe what happens" in the quantum-theoretical process between two successive observations is a contradiction *in adjecto,* since the word "describe" refers to the use of the classical concepts, while these concepts cannot be applied in the space between the observations; they can only be applied at the points of observation.

It should be noticed at this point that the Copenhagen interpretation of quantum theory is in no way positivistic. For, whereas positivism is based on the sensual perceptions of the observer as the elements of reality, the Copenhagen interpretation regards things and processes which are describable in terms of classical concepts, i.e., the actual, as the foundation of any physical interpretation.

At the same time we see that the statistical nature of the laws of microscopic physics cannot be avoided, since any knowledge of the "actual" is—because of the quantum-theoretical laws— by its very nature an incomplete knowledge.

The ontology of materialism rested upon the illusion that the kind of existence, the direct "actuality" of the world around us, can be extrapolated into the atomic range. This extrapolation is impossible, however.

A few remarks may be added concerning the formal structure of all the counterproposals hitherto made against the Copenhagen interpretation of quantum theory. All these proposals have found themselves compelled to sacrifice the essential symmetry properties of quantum theory (for instance, the symmetry between waves and particles or between position and velocity).

Therefore, we may well suppose that the Copenhagen interpretation cannot be avoided if these symmetry properties—like the Lorentz invariance in the theory of relativity—are held to be a genuine feature of nature; and every experiment yet performed supports this view.

IX.

Quantum Theory and the Structure of Matter

THE concept of matter has undergone a great number of changes in the history of human thinking. Different interpretations have been given in different philosophical systems. All these different meanings of the word are still present in a greater or lesser degree in what we conceive in our time as the word "matter."

The early Greek philosophy from Thales to the Atomists, in seeking the unifying principle in the universal mutability of all things, had formed the concept of cosmic matter, a world substance which experiences all these transformations, from which all individual things arise and into which they become again transformed. This matter was partly identified with some specific matter like water or air or fire; only partly, because it had no other attribute but to be the material from which all things are made.

Later, in the philosophy of Aristotle, matter was thought of in the relation between form and matter. All that we perceive in the world of phenomena around us is formed matter. Matter is in itself not a reality but only a possibility, a "potentia"; it exists only by means of form. In the natural process the "essence," as Aristotle calls it, passes over from mere possibility through form into actuality. The matter of Aristotle is certainly not a specific matter like water or air, nor is it simply empty space; it is a kind of indefinite corporeal substratum, embodying the possibility of passing over into actuality by means of the form. The typical examples of this relation between matter and form in the philosophy of Aristotle are the biological processes in which matter is formed to become the living organism, and the building and forming activity of man. The statue is potentially in the marble before it is cut out by the sculptor.

Then, much later, starting from the philosophy of Descartes, matter was primarily thought of as opposed to mind. There were the two complementary aspects of the world, "matter" and "mind," or, as Descartes put it, the "res extensa" and the "res cogitans." Since the new methodical principles of natural science, especially of mechanics, excluded all tracing of corporeal phenomena back to spiritual forces, matter could be considered as a reality of its own independent of the mind and of any supernatural powers. The "matter" of this period is "formed matter," the process of formation being interpreted as a causal chain of mechanical interactions; it has lost its connection with the vegetative soul of Aristotelian philosophy, and therefore the dualism between matter and form is no longer relevant. It is this concept of matter which constitutes by far the strongest component in our present use of the word "matter."

Finally, in the natural science of the nineteenth century another dualism has played some role, the dualism between matter and force. Matter is that on which forces can act; or matter can produce forces. Matter, for instance, produces the force of gravity, and this force acts on matter. Matter and force are two distinctly different aspects of the corporeal world. In so far as the forces may be formative forces this distinction comes closer to the Aristotelian distinction of matter and form. On the other hand, in the most recent development of modern physics this distinction between matter and force is completely lost, since every field of force contains energy and in so far constitutes matter. To every field of force there belongs a specific kind of elementary particles with essentially the same properties as all other atomic units of matter.

When natural science investigates the problem of matter it can do so only through a study of the forms of matter. The infinite variety and mutability of the forms of matter must be the immediate object of the investigation and the efforts must be directed toward finding some natural laws, some unifying principles that can serve as a guide through this immense field. Therefore, natural science—and especially physics—has concentrated its interest for a long period on an analysis of the structure of matter and of the forces responsible for this structure.

Since the time of Galileo the fundamental method of natural science had been the experiment. This method made it possible to pass from general experience to specific experience, to single out characteristic events in nature from which its "laws" could be studied more directly than from general experience. If one wanted to study the structure of matter one had to do experiments with matter. One had to expose matter to extreme conditions in order to study its transmutations there, in the hope of finding the fundamental features of matter which persist under all apparent changes.

In the early days of modern natural science this was the object of chemistry, and this endeavor led rather early to the concept of the chemical element. A substance that could not be further dissolved or disintegrated by any of the means at the disposal of the chemist—boiling, burning, dissolving, mixing with other substances, etc.—was called an element. The introduction of this concept was a first and most important step toward an understanding of the structure of matter. The enormous variety of substances was at least reduced to a comparatively small number of more fundamental substances, the "elements," and thereby some order could be established among the various phenomena of chemistry. The word "atom" was consequently used to designate the smallest unit of matter belonging to a chemical element, and the smallest particle of a chemical compound could be pictured as a small group of different atoms. The smallest particle of the element iron, e.g., was an iron atom, and the smallest particle of water, the water molecule, consisted of one oxygen atom and two hydrogen atoms.

The next and almost equally important step was the discovery of the conservation of mass in the chemical process. For instance, when the element carbon is burned into carbon dioxide the mass of the carbon dioxide is equal to the sum of the masses of the carbon and the oxygen before the process. It was this discovery that gave a quantitative meaning to the concept of matter: independent of its chemical properties matter could be measured by its mass.

During the following period, mainly the nineteenth century, a number of new chemical elements were discovered; in our time this number has reached one hundred. This development showed quite clearly that the concept of the chemical element had not yet reached the point where one could understand the unity of matter. It was not satisfactory to believe that there are very many kinds of matter, qualitatively different and without any connection between one another.

In the beginning of the nineteenth century some evidence for a connection between the different elements was found in the fact that the atomic weights of different elements frequently seemed to be integer multiples of a smallest unit near to the atomic weight of hydrogen. The similarity in the chemical behavior of some elements was another hint leading in the same direction. But only the discovery of forces much stronger than those applied in chemical processes could really establish the connection between the different elements and thereby lead to a closer unification of matter.

These forces were actually found in the radioactive process discovered in 1896 by Becquerel. Successive investigations by Curie, Rutherford and others revealed the transmutation of elements in the radioactive process. The α-particles are emitted in these processes as fragments of the atoms with an energy about a million times greater than the energy of a single atomic particle in a chemical process. Therefore, these particles could be used as new tools for investigating the inner structure of the atom. The result of Rutherford's experiments on the scattering of α-rays was the nuclear model of the atom in 1911. The most important feature of this well-known model was the separation of the atom into two distinctly different parts, the atomic nucleus and the surrounding electronic shells. The nucleus in the middle of the atom occupies only an extremely small fraction of the space filled by the atom (its radius is about a hundred thousand times smaller than that of the atom), but contains almost its entire mass. Its positive electric charge, which is an integer multiple of the so-called elementary charge, determines the number of the surrounding electrons—the atom as a whole must be electrically neutral—and the shapes of their orbits.

This distinction between the atomic nucleus and the electronic shells at once gave a proper explanation of the fact that for chemistry the chemical elements are the last units of matter and that very much stronger forces are required to change the elements into each other. The chemical bond between neighboring atoms is due to an interaction of the electronic shells, and the energies of this interaction are comparatively small. An electron that is accelerated in a discharge tube by a potential of only several volts has sufficient energy to excite the electronic shells to the emission of radiation, or to destroy the chemical bond in a molecule. But the chemical behavior of the atom, though it consists of the behavior of its electronic shells, is determined by the charge of the nucleus. One has to change the nucleus if one wants to change the chemical properties, and this requires energies about a million times greater.

The nuclear model of the atom, however, if it is thought of as a system obeying Newton's mechanics, could not explain the stability of the atom. As has been pointed out in an earlier chapter, only the application of quantum theory to this model through the work of Bohr could account for the fact that, for example, a carbon atom after having been in interaction with other atoms or after having emitted radiation always finally remains a carbon atom with the same electronic shells as before. This stability could be explained simply by those features of quantum theory that prevent a simple objective description in space and time of the structure of the atom.

In this way one finally had a first basis for the understanding of matter. The chemical and other properties of the atoms could be accounted for by applying the mathematical scheme of quantum theory to the electronic shells. From this basis one could try to extend the analysis of the structure of matter in two opposite directions. One could either study the interaction of atoms, their relation to larger units like molecules or crystals or biological objects; or one could try through the investigation of the atomic nucleus and its components to penetrate to the final unity of matter. Research has proceeded on both lines during the past decades and we shall in the following pages be concerned with the role of quantum theory in these two fields.

The forces between neighboring atoms are primarily electric forces, the attraction of opposite and the repulsion of equal charges; the electrons are attracted by the nuclei and repelled from each other. But these forces act not according to the laws of Newtonian mechanics but those of quantum mechanics.

This leads to two different types of binding between atoms. In the one type the electron of one atom passes over to the other one, for example, to fill up a nearly closed electronic shell. In this case both atoms are finally charged and form what the physicist calls ions, and since their charges are opposite they attract each other.

In the second type one electron belongs in a way characteristic of quantum theory to both atoms. Using the picture of the electronic orbit, one might say that the electron goes around both nuclei spending a comparable amount of time in the one and in the other atom. This second type of binding corresponds to what the chemists call a valency bond.

These two types of forces, which may occur in any mixture, cause the formation of various groupings of atoms and seem to be ultimately responsible for all the complicated structures of matter in bulk that are studied in physics and chemistry. The formation of chemical compounds takes place through the formation of small closed groups of different atoms, each group being one molecule of the compound. The formation of crystals is due to the arrangement of the atoms in regular lattices. Metals are formed when the atoms are so tightly packed that their outer electrons can leave their shells and wander through the whole crystal. Magnetism is due to the spinning motion of the electron, and so on.

In all these cases the dualism between matter and force can still be retained, since one may consider nuclei and electrons as the fragments of matter that are kept together by means of the electromagnetic forces.

While in this way physics and chemistry have come to an almost complete union in their relations to the structure of matter, biology deals with structures of a more complicated and somewhat different type. It is true that in spite of the wholeness of the living organism a sharp distinction between animate and inanimate matter can certainly not be made. The development of biology has supplied us with a great number of examples where one can see that specific biological functions are carried by special large molecules or groups or chains of such molecules, and there has been an increasing tendency in modern biology to explain biological processes as consequences of the laws of physics and chemistry. But the kind of stability that is displayed by the living organism is of a nature somewhat different from the stability of atoms or crystals. It is a stability of process or function rather than a stability of form. There can be no doubt that the laws of quantum theory play a very important role in the biological phenomena. For instance, those specific quantum-theoretical forces that can be described only inaccurately by the concept of chemical valency are essential for the understanding of the big organic molecules and their various geometrical patterns; the experiments on biological mutations produced by radiation show both the relevance of the statistical quantum-theoretical laws and the existence of amplifying mechanisms. The close analogy between the working of our nervous system and the functioning of modern electronic computers stresses again the importance of single elementary processes in the living organism. Still all this does not prove that physics and chemistry will, together with the concept of evolution, someday offer a complete description of the living organism. The biological processes must be handled by the experimenting scientist with greater caution than processes of physics and chemistry. As Bohr has pointed out, it may well be that a description of the living organism that could be called complete from the standpoint of the physicist cannot be given, since it would require experiments that interfere too strongly with the biological functions. Bohr has described this situation by saying that in biology we are concerned with manifestations of possibilities in that nature to which we belong rather than with outcomes of experiments which we can ourselves perform. The situation of complementarity to which this formulation alludes is represented as a tendency in the methods of modern biological research which, on the one hand, makes full use of all the methods and results of physics and chemistry and, on the other hand, is based on concepts referring to those features of organic nature that are not contained in physics or chemistry, like the concept of life itself.

So far we have followed the analysis of the structure of matter in one direction: from the atom to the more complicated structures consisting of many atoms; from atomic physics to the physics of solid bodies, to chemistry and to biology. Now we have to turn to the opposite direction and follow the line of research from the outer parts of the atom to the inner parts and from the nucleus to the elementary particles. It is this line which will possibly lead to an understanding of the unity of matter. Here we need not be afraid of destroying characteristic structures by our experiments. When the task is set to test the final unity of matter we may expose matter to the strongest possible forces, to the most extreme conditions, in order to see whether any matter can ultimately be transmuted into any other matter.

The first step in this direction was the experimental analysis of the atomic nucleus. In the initial period of these studies, which filled approximately the first three decades of our century, the only tools available for experiments on the nucleus were the α-particles emitted from radioactive bodies. With the help of these particles Rutherford succeeded in 1919 in transmuting nuclei of light elements; he could, for instance, transmute a nitrogen nucleus into an oxygen nucleus by adding the α-particle to the nitrogen nucleus and at the same time knocking out one proton. This was the first example of processes on a nuclear scale that reminded one of chemical processes, but led to the artificial transmutation of elements. The next substantial progress was, as is well known, the artificial acceleration of protons by means of high-tension equipment to energies sufficient to cause nuclear transmutation. Voltages of roughly one million volts are required for this purpose and Cockcroft and Walton in their first decisive experiment succeeded in transmuting nuclei of the element lithium into those of helium. This discovery opened up an entirely new line of research, which may be called nuclear physics in the proper sense and which very soon led to a qualitative understanding of the structure of the atomic nucleus.

The structure of the nucleus was indeed very simple. The atomic nucleus consists of only two kinds of elementary particles. The one is the proton which is at the same time simply the hydrogen nucleus; the other is called neutron, a particle which has roughly the mass of the proton but is electrically neutral. Every nucleus can be characterized by the number of protons and neutrons of which it consists. The normal carbon nucleus, for instance, consists of 6 protons and 6 neutrons. There are other carbon nuclei, less frequent in number (called isotopic to the first ones), that consist of 6 protons and 7 neutrons, etc. So one had finally reached a description of matter in which, instead of the many different chemical elements, only three fundamental units occurred: the proton, the neutron and the electron. All matter consists of atoms and therefore is constructed from these three fundamental building stones. This was not yet the unity of matter, but certainly a great step toward unification and—perhaps still more important—simplification. There was of course still a long way to go from the knowledge of the two building stones of the nucleus to a complete understanding of its structure. The problem here was somewhat different from the corresponding problem in the outer atomic shells that had been solved in the middle of the twenties. In the electronic shells the forces between the particles were known with great accuracy, but the dynamic laws had to be found, and were found in quantum mechanics. In the nucleus the dynamic laws could well be supposed to be just those of quantum mechanics, but the forces between the particles were not known beforehand; they had to be derived from the experimental properties of the nuclei. This problem has not yet been completely solved. The forces have probably not such a simple form as the electrostatic forces in the electronic shells and therefore the mathematical difficulty of computing the properties from complicated forces and the inaccuracy of the experiments make progress difficult. But a qualitative understanding of the structure of the nucleus has definitely been reached.

Then there remained the final problem, the unity of matter. Are these fundamental building stones—proton, neutron and electron—final indestructible units of matter, atoms in the sense of Democritus, without any relation except for the forces that act between them or are they just different forms of the same kind of matter? Can they again be transmuted into each other and possibly into other forms of matter as well? An experimental attack on this problem requires forces and energies concentrated on atomic particles much larger than those that have been necessary to investigate the atomic nucleus. Since the energies stored up in atomic nuclei are not big enough to provide us with a tool for such experiments, the physicists have to rely either on the forces in cosmic dimensions or on the ingenuity and skill of the engineers.

Actually, progress has been made on both lines. In the first case the physicists make use of the so-called cosmic radiation. The electomagnetic fields on the surface of stars extending over huge spaces are under certain circumstances able to accelerate charged atomic particles, electrons and nuclei. The nuclei, owing to their greater inertia, seem to have a better chance of remaining in the accelerating field for a long distance, and finally when they leave the surface of the star into empty space they have already traveled through potentials of several thousand million volts. There may be a further acceleration in the magnetic fields between the stars; in any case the nuclei seem to be kept within the space of the galaxy for a long time by varying magnetic fields, and finally they fill this space with what one calls cosmic radiation. This radiation reaches the earth from the outside and consists of nuclei of practically all kinds, hydrogen and helium and many heavier elements, having energies from roughly a hundred or a thousand million electron volts to, again in rare cases, a million times this amount. When the particles of this cosmic radiation penetrate into the atmosphere of the earth they hit the nitrogen atoms or oxygen atoms of the atmosphere or may hit the atoms in any experimental equipment exposed to the radiation.

The other line of research was the construction of big accelerating machines, the prototype of which was the so-called cyclotron constructed by Lawrence in California in the early thirties. The underlying idea of these machines is to keep by means of a big magnetic field the charged particles going round in circles a great number of times so that they can be pushed again and again by electric fields on their way around. Machines reaching up to energies of several hundred million electron volts are in use in Great Britain, and through the co-operation of twelve European countries a very big machine of this type is now being constructed in Geneva which we hope will reach up to energies of 25,000 million electron volts. The experiments carried out by means of cosmic radiation or of the big accelerators have revealed new interesting features of matter. Besides the three fundamental building stones of matter—electron, proton and neutron—new elementary particles have been found which can be created in these processes of highest energies and disappear again after a short time. The new particles have similar properties as the old ones except for their instability. Even the most stable ones have lifetimes of roughly only a millionth part of a second, and the lifetimes of others are even a thousand times smaller. At the present time about twenty-five different new elementary particles are known; the most recent one is the negative proton.

These results seem at first sight to lead away from the idea of the unity of matter, since the number of fundamental units of matter seems to have again increased to values comparable to the number of different chemical elements. But this would not be a proper interpretation. The experiments have at the same time shown that the particles can be created from other particles or simply from the kinetic energy of such particles, and they can again disintegrate into other particles. Actually the experiments have shown the complete mutability of matter. All the elementary particles can, at sufficiently high energies, be transmuted into other particles, or they can simply be created from kinetic energy and can be annihilated into energy, for instance into radiation. Therefore, we have here actually the final proof for the unity of matter. All the elementary particles are made of the same substance, which we may call energy or universal matter; they are just different forms in which matter can appear.

If we compare this situation with the Aristotelian concepts of matter and form, we can say that the matter of Aristotle, which is mere "potentia," should be compared to our concept of energy, which gets into "actuality" by means of the form, when the elementary particle is created.

Modern physics is of course not satisfied with only qualitative description of the fundamental structure of matter; it must try on the basis of careful experimental investigations to get a mathematical formulation of those natural laws that determine the "forms" of matter, the elementary particles and their forces. A clear distinction between matter and force can no longer be made in this part of physics, since each elementary particle not only is producing some forces and is acted upon by forces, but it is at the same time representing a certain field of force. The quantum-theoretical dualism of waves and particles makes the same entity appear both as matter and as force.

All the attempts to find a mathematical description for the laws concerning the elementary particles have so far started from the quantum theory of wave fields. Theoretical work on theories of this type started early in the thirties. But the very first investigations on this line revealed serious difficulties the roots of which lay in the combination of quantum theory and the theory of special relativity. At first sight it would seem that the two theories, quantum theory and the theory of relativity, refer to such different aspects of nature that they should have practically nothing to do with each other, that it should be easy to fulfill the requirements of both theories in the same formalism. A closer inspection, however, shows that the two theories do interfere at one point, and that it is from this point that all the difficulties arise.

The theory of special relativity had revealed a structure of space and time somewhat different from the structure that was generally assumed since Newtonian mechanics. The most characteristic feature of this newly discovered structure is the existence of a maximum velocity that cannot be surpassed by any moving body or any traveling signal, the velocity of light. As a consequence of this, two events at distant points cannot have any immediate causal connection if they take place at such times that a light signal starting at the instant of the event on *one* point reaches the other point only after the time the other event has happened there; and vice versa. In this case the two events may be called simultaneous. Since no action of any kind can reach from the one event at the one point in time to the other event at the other point, the two events are not connected by any causal action.

For this reason any action at a distance of the type, say, of the gravitational forces in Newtonian mechanics was not compatible with the theory of special relativity. The theory had to replace such action by actions from point to point, from one point only to the points in an infinitesimal neighborhood. The most natural mathematical expressions for actions of this type were the differential equations for waves or fields that were invariant for the Lorentz transformation. Such differential equations exclude any direct action between "simultaneous" events.

Therefore, the structure of space and time expressed in the theory of special relativity implied an infinitely sharp boundary between the region of simultaneousness, in which no action could be transmitted, and the other regions, in which a direct action from event to event could take place.

On the other hand, in quantum theory the uncertainty relations put a definite limit on the accuracy with which positions and momenta, or time and energy, can be measured simultaneously. Since an infinitely sharp boundary means an infinite accuracy with respect to position in space and time, the momenta or energies must be completely undetermined, or in fact arbitrarily high momenta and energies must occur with overwhelming probability. Therefore, any theory which tries to fulfill the requirements of both special relativity and quantum theory will lead to mathematical inconsistencies, to divergencies in the region of very high energies and momenta. This sequence of conclusions may perhaps not seem strictly binding, since any formalism of the type under consideration is very complicated and could perhaps offer some mathematical possibilities for avoiding the clash between quantum theory and relativity. But so far all the mathematical schemes that have been tried did in fact lead either to divergencies, i.e., to mathematical contradictions, or did not fulfill all the requirements of the two theories. And it was easy to see that the difficulties actually came from the point that has been discussed.

The way in which the convergent mathematical schemes did not fulfill the requirements of relativity or quantum theory was in itself quite interesting. For instance, one scheme, when interpreted in terms of actual events in space and time, led to a kind of time reversal; it would predict processes in which suddenly at some point in space particles are created, the energy of which is later provided for by some other collision process between elementary particles at some other point. The physicists are convinced from their experiments that processes of this type do not occur in nature, at least not if the two processes are separated by measurable distances in space and time. Another mathematical scheme tried to avoid the divergencies through a mathematical process which is called renormalization; it seemed possible to push the infinities to a place in the formalism where they could not interfere with the establishment of the well-defined relations between those quantities that can be directly observed. Actually this scheme has led to very substantial progress in quantum electrodynamics, since it accounts for some interesting details in the hydrogen spectrum that had not been understood before. A closer analysis of this mathematical scheme, however, has made it probable that those quantities which in normal quantum theory must be interpreted as probabilities can under certain circumstances become negative in the formalism of renormalization. This would prevent the consistent use of the formalism for the description of matter.

The final solution of these difficulties has not yet been found. It will emerge someday from the collection of more and more accurate experimental material about the different elementary particles, their creation and annihilation, the forces between them. In looking for possible solutions of the difficulties one should perhaps remember that such processes with time reversal as have been discussed before could not be excluded experimentally, if they took place only within extremely small regions of space and time outside the range of our present experimental equipment. Of course one would be reluctant to accept such processes with time reversal if there could be at any later stage of physics the possibility of following experimentally such events in the same sense as one follows ordinary atomic events. But here the analysis of quantum theory and of relativity may again help us to see the problem in a new light.

The theory of relativity is connected with a universal constant in nature, the velocity of light. This constant determines the relation between space and time and is therefore implicitly contained in any natural law which must fulfill the requirements of Lorentz invariance. Our natural language and the concepts of classical physics can apply only to phenomena for which the velocity of light can be considered as practically infinite.

When we in our experiments approach the velocity of light we must be prepared for results which cannot be interpreted in these concepts.

Quantum theory is connected with another universal constant of nature, Planck's quantum of action. An objective description for events in space and time is possible only when we have to deal with objects or processes on a comparatively large scale, where Planck's constant can be regarded as infinitely small. When our experiments approach the region where the quantum of action becomes essential we get into all those difficulties with the usual concepts that have been discussed in earlier chapters of this volume.

There must exist a third universal constant in nature. This is obvious for purely dimensional reasons. The universal constants determine the scale of nature, the characteristic quantities that cannot be reduced to other quantities. One needs at least three fundamental units for a complete set of units. This is most easily seen from such conventions as the use of the c-g-s system (centimeter, gram, second system) by the physicists. A unit of length, one of time, and one of mass is sufficient to form a complete set; but one must have at least three units. One could also replace them by units of length, velocity and mass; or by units of length, velocity and energy, etc. But at least three fundamental units are necessary. Now, the velocity of light and Planck's constant of action provide only two of these units. There must be a third one, and only a theory which contains this third unit can possibly determine the masses and other properties of the elementary particles. Judging from our present knowledge of these particles the most appropriate way of introducing the third universal constant would be by the assumption of a universal length the value of which should be roughly 10^{-13} cm, that is, somewhat smaller than the radii of the light atomic nuclei. When from such three units one forms an expression which in its dimension corresponds to a mass, its value has the order of magnitude of the masses of the elementary particles.

If we assume that the laws of nature *do* contain a third universal constant of the dimension of a length and of the order of 10^{-13} cm, then we would again expect our usual concepts to apply only to regions in space and time that are large as compared to the universal constant. We should again be prepared for phenomena of a qualitatively new character when we in our experiments approach regions in space and time smaller than the nuclear radii. The phenomenon of time reversal, which has been discussed and which so far has only resulted from theoretical considerations as a mathematical possibility, might therefore belong to these smallest regions. If so, it could probably not be observed in a way that would permit a description in terms of the classical concepts. Such processes would probably, so far as they can be observed and described in classical terms, obey the usual time order.

But all these problems will be a matter of future research in atomic physics. One may hope that the combined effort of experiments in the high energy region and of mathematical analysis will someday lead to a complete understanding of the unity of matter. The term "complete understanding" would mean that the forms of matter in the sense of Aristotelian philosophy would appear as results, as solutions of a closed mathematical scheme representing the natural laws for matter.

X.

Language and Reality in Modern Physics

THROUGHOUT the history of science new discoveries and new ideas have always caused scientific disputes, have led to polemical publications criticizing the new ideas, and such criticism has often been helpful in their development; but these controversies have never before reached that degree of violence which they attained after the discovery of the theory of relativity and in a lesser degree after quantum theory. In both cases the scientific problems have finally become connected with political issues, and some scientists have taken recourse to political methods to carry their views through. This violent reaction on the recent development of modern physics can only be understood when one realizes that here the foundations of physics have started moving; and that this motion has caused the feeling that the ground would be cut from science. At the same time it probably means that one has not yet found the correct language with which to speak about the new situation and that the incorrect statements published here and there in the enthusiasm about the new discoveries have caused all kinds of misunderstanding. This is indeed a fundamental problem. The improved experimental technique of our time brings into the scope of science new aspects of nature which cannot be described in terms of the common concepts. But in what language, then, should they be described? The first language that emerges from the process of scientific clarification is in theoretical physics usually a mathematical language, the mathematical scheme, which allows one to predict the results of experiments. The physicist may be satisfied when he has the mathematical scheme and knows how to use it for the interpretation of the experiments. But he has to speak about his results also to nonphysicists who will not be satisfied unless some explanation is given in plain language, understandable to anybody. Even for the physicist the description in plain language will be a criterion of the degree of understanding that has been reached. To what extent is such a description at all possible? Can one speak about the atom itself? This is a problem of language as much as of physics, and therefore some remarks are necessary concerning language in general and scientific language specifically.

Language was formed during the prehistoric age among the human race as a means for communication and as a basis for thinking. We know little about the various steps in its formation; but language now contains a great number of concepts which are a suitable tool for more or less unambiguous communication about events in daily life. These concepts are acquired gradually without critical analysis by using the language, and after having used a word sufficiently often we think that we more or less know what it means. It is of course a well-known fact that the words are not so clearly defined as they seem to be at first sight and that they have only a limited range of applicability. For instance, we can speak about a piece of iron or a piece of wood, but we cannot speak about a piece of water. The word "piece" does not apply to liquid substances. Or, to mention another example: In discussions about the limitations of concepts, Bohr likes to tell the following story: "A little boy goes into a grocer's shop with a penny in his hand and asks: 'Could I have a penny's worth of mixed sweets?' The grocer takes two sweets and hands them to the boy saying: 'Here you have two sweets. You can do the mixing yourself.' " A more serious example of the problematic relation between words and concepts is the fact that the words "red" and "green" are used even by people who are colorblind, though the ranges of applicability of these terms must be quite different for them from what they are for other people.

This intrinsic uncertainty of the meaning of words was of course recognized very early and has brought about the need for definitions, or—as the word "definition" says—for the setting of boundaries that determine where the word is to be used and where not. But definitions can be given only with the help of other concepts, and so one will finally have to rely on some concepts that are taken as they are, unanalyzed and undefined.

In Greek philosophy the problem of the concepts in language has been a major theme since Socrates, whose life was—if we can follow Plato's artistic representation in his dialogues—a continuous discussion about the content of the concepts in language and about the limitations in modes of expression. In order to obtain a solid basis for scientific thinking, Aristotle in his logic started to analyze the forms of language, the formal structure of conclusions and deductions independent of their content. In this way he reached a degree of abstraction and precision that had been unknown up to that time in Greek philosophy and he thereby contributed immensely to the clarification, to the establishment of order in our methods of thought. He actually created the basis for the scientific language.

On the other hand, this logical analysis of language again involves the danger of an oversimplification. In logic the attention is drawn to very special structures, unambiguous connections between premises and deductions, simple patterns of reasoning, and all the other structures of language are neglected. These other structures may arise from associations between certain meanings of words; for instance, a secondary meaning of a word which passes only vaguely through the mind when the word is heard may contribute essentially to the content of a sentence. The fact that every word may cause many only half-conscious movements in our mind can be used to represent some part of reality in the language much more clearly than by the use of the logical patterns. Therefore, the poets have often objected to this emphasis in language and in thinking on the logical pattern, which—if I interpret their opinions correctly—can make language less suitable for its purpose. We may recall for instance the words in Goethe's *Faust* which Mephistopheles speaks to the young student (quoted from the translation by Anna Swanwick):

> Waste not your time, so fast it flies;
> Method will teach you time to win;
> Hence, my young friend, I would advise,
> With college logic to begin.
> Then will your mind be so well brac'd,
> In Spanish boots so tightly lac'd,
> That on 'twill circumspectly creep,
> Thought's beaten track securely keep,
> Nor will it, ignis-fatuus like,
> Into the path of error strike.
> Then many a day they'll teach you how
> The mind's spontaneous acts, till now

As eating and as drinking free,
Require a process;—one, two, three!
In truth the subtle web of thought
Is like the weaver's fabric wrought,
One treadle moves a thousand lines,
Swift dart the shuttles to and fro,
Unseen the threads unnumber'd flow,
A thousand knots one stroke combines.
Then forward steps your sage to show,
And prove to you it must be so;
The first being so, and so the second.
The third and fourth deduc'd we see;
And if there were no first and second,
Nor third nor fourth would ever be.
This, scholars of all countries prize,
Yet 'mong themselves no weavers rise.
Who would describe and study aught alive,
Seeks first the living spirit thence to drive:
Then are the lifeless fragments in his hand,
There only fails, alas!—the spirit-band.

This passage contains an admirable description of the structure of language and of the narrowness of the simple logical patterns.

On the other hand, science must be based upon language as the only means of communication and there, where the problem of unambiguity is of greatest importance, the logical patterns must play their role. The characteristic difficulty at this point may be described in the following way. In natural science we try to derive the particular from the general, to understand the particular phenomenon as caused by simple general laws. The general laws when formulated in the language can contain only a few simple concepts—else the law would not be simple and general. From these concepts are derived an infinite variety of possible phenomena, not only qualitatively but with complete precision with respect to every detail. It is obvious that the concepts of ordinary language, inaccurate and only vaguely defined as they are, could never allow such derivations. When a chain of conclusions follows from given premises, the number of possible links in the chain depends on the precision of the premises. Therefore, the concepts of the general laws must in natural science be defined with complete precision, and this can be achieved only by means of mathematical abstraction.

In other sciences the situation may be somewhat similar in so far as rather precise definitions are also required; for instance, in law. But here the number of links in the chain of conclusions need not be very great, complete precision is not needed, and rather precise definitions in terms of ordinary language are sufficient.

In theoretical physics we try to understand groups of phenomena by introducing mathematical symbols that can be correlated with facts, namely, with the results of measurements. For the symbols we use names that visualize their correlation with the measurement. Thus the symbols are attached to the language. Then the symbols are interconnected by a rigorous system of definitions and axioms, and finally the natural laws are expressed as equations between the symbols. The infinite variety of solutions of these equations then corresponds to the infinite variety of particular phenomena that are possible in this part of nature. In this way the mathematical scheme represents the group of phenomena so far as the correlation between the symbols and the measurements goes. It is this correlation which permits the expression of natural laws in the terms of common language, since our experiments consisting of actions and observations can always be described in ordinary language.

Still, in the process of expansion of scientific knowledge the language also expands; new terms are introduced and the old ones are applied in a wider field or differently from ordinary language. Terms such as "energy," "electricity," "entropy" are obvious examples. In this way we develop a scientific language which may be called a natural extension of ordinary language adapted to the added fields of scientific knowledge.

During the past century a number of new concepts have been introduced in physics, and in some cases it has taken considerable time before the scientists have really grown accustomed to their use. The term "electromagnetic field," for instance, which was to some extent already present in Faraday's work and which later formed the basis of Maxwell's theory, was not easily accepted by the physicists, who directed their attention primarily to the mechanical motion of matter. The introduction of the concept really involved a change in scientific ideas as well, and such changes are not easily accomplished.

Still, all the concepts introduced up to the end of the last century formed a perfectly consistent set applicable to a wide field of experience, and, together with the former concepts, formed a language which not only the scientists but also the technicians and engineers could successfully apply in their work. To the underlying fundamental ideas of this language belonged the assumptions that the order of events in time is entirely independent of their order in space, that Euclidean geometry is valid in real space, and that the events "happen" in space and time independently of whether they are observed or not. It was not denied that every observation had some influence on the phenomenon to be observed but it was generally assumed that by doing the experiments cautiously this influence could be made arbitrarily small. This seemed in fact a necessary condition for the ideal of objectivity which was considered as the basis of all natural science.

Into this rather peaceful state of physics broke quantum theory and the theory of special relativity as a sudden, at first slow and then gradually increasing, movement in the foundations of natural science. The first violent discussions developed around the problems of space and time raised by the theory of relativity. How should one speak about the new situation? Should one consider the Lorentz contraction of moving bodies as a real contraction or only as an apparent contraction? Should one say that the structure of space and time was really different from what it had been assumed to be or should one only say that the experimental results could be connected mathematically in a way corresponding to this new structure, while space and time, being the universal and necessary mode in which things appear to us, remain what they had always been? The real problem behind these many controversies was the fact that no language existed in which one could speak consistently about the new situation. The ordinary language was based upon the old concepts of space and time and this language offered the only unambiguous means of communication about the setting up and the results of the measurements. Yet the experiments showed that the old concepts could not be applied everywhere.

The obvious starting point for the interpretation of the theory of relativity was therefore the fact that in the limiting case of small velocities (small compared with the velocity of light) the new theory was practically identical with the old one. Therefore, in this part of the theory it was obvious in which way the mathematical symbols had to be correlated with the measurements and with the terms of ordinary language; actually it was only through this correlation that the Lorentz transformation had been found. There was no ambiguity about the meaning of the words and the symbols in this region. In fact this correlation was already sufficient for the application of the theory to the whole field of experimental research connected with the problem of relativity. Therefore, the controversial questions about the "real" or the "apparent" Lorentz contraction, or about the definition of the word "simultaneous" etc., did not concern the facts but rather the language.

With regard to the language, on the other hand, one has gradually recognized that one should perhaps not insist too much on certain principles. It is always difficult to find general convincing criteria for which terms should be used in the language and how they should be used. One should simply wait for the development of the language, which adjusts itself after some time to the new situation. Actually in the theory of special relativity this adjustment has already taken place to a large extent during the past fifty years. The distinction between "real" and "apparent" contraction, for instance, has simply disappeared. The word "simultaneous" is used in line with the definition given by Einstein, while for the wider definition discussed in an earlier chapter the term "at a space-like distance" is commonly used, etc.

In the theory of general relativity the idea of a non-Euclidean geometry in real space was strongly contradicted by some philosophers who pointed out that our whole method of setting up the experiments already presupposed Euclidean geometry.

In fact if a mechanic tries to prepare a perfectly plane surface, he can do it in the following way. He first prepares three surfaces of, roughly, the same size which are, roughly, plane. Then he tries to bring any two of the three surfaces into contact by putting them against each other in different relative positions. The degree to which this contact is possible on the whole surface is a measure of the degree of accuracy with which the surfaces can be called "plane." He will be satisfied with his three surfaces only if the contact between any two of them is complete everywhere. If this happens one can prove mathematically that Euclidean geometry holds on the three surfaces. In this way, it was argued, Euclidean geometry is just *made* correct by our own measures.

From the point of view of general relativity, of course, one can answer that this argument proves the validity of Euclidean geometry only in small dimensons, in the dimensions of our experimental equipment. The accuracy with which it holds in this region is so high that the above process for getting plane surfaces can always be carried out. The extremely slight deviations from Euclidean geometry which still exist in this region will not be realized since the surfaces are made of material which is not strictly rigid but allows for very small deformations and since the concept of "contact" cannot be defined with complete precision. For surfaces on a cosmic scale the process that has been described would just not work; but this is not a problem of experimental physics.

Again, the obvious starting point for the physical interpretation of the mathematical scheme in general relativity is the fact that the geometry is very nearly Euclidean in small dimensions; the theory approaches the classical theory in this region. Therefore, here the correlation between the mathematical symbols and the measurements and the concepts in ordinary language is unambiguous. Still, one can speak about a non-Euclidean geometry in large dimensions. In fact a long time before the theory of general relativity had even been developed the possibility of a non-Euclidean geometry in real space seems to have been considered by the mathematicians, especially by Gauss in Göttingen. When he carried out very accurate geodetic measurements on a triangle formed by three mountains—the Brocken in the Harz Mountains, the Inselberg in Thuringia, and the Hohenhagen near Göttingen—he is said to have checked very carefully whether the sum of the three angles was actually equal to 180 degrees; and that he considered a difference which would prove deviations from Euclidean geometry as being possible. Actually he did not find any deviations within his accuracy of measurement.

In the theory of general relativity the language by which we describe the general laws actually now follows the scientific language of the mathematicians, and for the description of the experiments themselves we can use the ordinary concepts, since Euclidean geometry is valid with sufficient accuracy in small dimensions.

The most difficult problem, however, concerning the use of the language arises in quantum theory. Here we have at first no simple guide for correlating the mathematical symbols with concepts of ordinary language; and the only thing we know from the start is the fact that our common concepts cannot be applied to the structure of the atoms. Again the obvious starting point for the physical interpretation of the formalism seems to be the fact that the mathematical scheme of quantum mechanics approaches that of classical mechanics in dimensions which are large as compared to the size of the atoms. But even this statement must be made with some reservations. Even in large dimensions there are many solutions of the quantum-theoretical equations to which no analogous solutions can be found in classical physics. In these solutions the phenomenon of the "interference of probabilities" would show up, as was discussed in the earlier chapters;

it does not exist in classical physics. Therefore, even in the limit of large dimensions the correlation between the mathematical symbols, the measurements, and the ordinary concepts is by no means trivial. In order to get to such an unambiguous correlation one must take another feature of the problem into account. It must be observed that the system which is treated by the methods of quantum mechanics is in fact a part of a much bigger system (eventually the whole world); it is interacting with this bigger system; and one must add that the microscopic properties of the bigger system are (at least to a large extent) unknown. This statement is undoubtedly a correct description of the actual situation. Since the system could not be the object of measurements and of theoretical investigations, it would in fact not belong to the world of phenomena if it had no interactions with such a bigger system of which the observer is a part. The interaction with the bigger system with its undefined microscopic properties then introduces a new statistical element into the description—both the quantum-theoretical and the classical one—of the system under consideration. In the limiting case of the large dimensions this statistical element destroys the effects of the "interference of probabilities" in such a manner that now the quantum-mechanical scheme really approaches the classical one in the limit. Therefore, at this point the correlation between the mathematical symbols of quantum theory and the concepts of ordinary language is unambiguous, and this correlation suffices for the interpretation of the experiments. The remaining problems again concern the language rather than the facts, since it belongs to the concept "fact" that it can be described in ordinary language.

But the problems of language here are really serious. We wish to speak in some way about the structure of the atoms and not only about the "facts"—the latter being, for instance, the black spots on a photographic plate or the water droplets in a cloud chamber. But we cannot speak about the atoms in ordinary language.

The analysis can now be carried further in two entirely different ways. We can either ask which language concerning the atoms has actually developed among the physicists in the thirty years that have elapsed since the formulation of quantum mechanics. Or we can describe the attempts for defining a precise scientific language that corresponds to the mathematical scheme.

In answer to the first question one may say that the concept of complementarity introduced by Bohr into the interpretation of quantum theory has encouraged the physicists to use an ambiguous rather than an unambiguous language, to use the classical concepts in a somewhat vague manner in conformity with the principle of uncertainty, to apply alternatively different classical concepts which would lead to contradictions if used simultaneously. In this way one speaks about electronic orbits, about matter waves and charge density, about energy and momentum, etc., always conscious of the fact that these concepts have only a very limited range of applicability. When this vague and unsystematic use of the language leads into difficulties, the physicist has to withdraw into the mathematical scheme and its unambiguous correlation with the experimental facts.

This use of the language is in many ways quite satisfactory, since it reminds us of a similar use of the language in daily life or in poetry. We realize that the situation of complementarity is not confined to the atomic world alone; we meet it when we reflect about a decision and the motives for our decision or when we have the choice between enjoying music and analyzing its structure. On the other hand, when the classical concepts are used in this manner, they always retain a certain vagueness, they acquire in their relation to "reality" only the same statistical significance as the concepts of classical thermodynamics in its statistical interpretation. Therefore, a short discussion of these statistical concepts of thermodynamics may be useful.

The concept "temperature" in classical thermodynamics seems to describe an objective feature of reality, an objective property of matter. In daily life it is quite easy to define with the help of a thermometer what we mean by stating that a piece of matter has a certain temperature. But when we try to define what the temperature of an atom could mean we are, even in classical physics, in a much more difficult position. Actually we cannot correlate this concept "temperature of the atom" with a well-defined property of the atom but have to connect it at least partly with our insufficient knowledge of it. We can correlate the value of the temperature with certain statistical expectations about the properties of the atom, but it seems rather doubtful whether an expectation should be called objective. The concept "temperature of the atom" is not much better defined than the concept "mixing" in the story about the boy who bought mixed sweets.

In a similar way in quantum theory all the classical concepts are, when applied to the atom, just as well and just as little defined as the "temperature of the atom"; they are correlated with statistical expectations; only in rare cases may the expectation become the equivalent of certainty. Again, as in classical thermodynamics, it is difficult to call the expectation objective. One might perhaps call it an objective tendency or possibility, a "potentia" in the sense of Aristotelian philosophy. In fact, I believe that the language actually used by physicists when they speak about atomic events produces in their minds similar notions as the concept "potentia." So the physicists have gradually become accustomed to considering the electronic orbits, etc., not as reality but rather as a kind of "potentia." The language has already adjusted itself, at least to some extent, to this true situation. But it is not a precise language in which one could use the normal logical patterns; it is a language that produces pictures in our mind, but together with them the notion that the pictures have only a vague connection with reality, that they represent only a tendency toward reality.

The vagueness of this language in use among the physicists has therefore led to attempts to define a different precise language which follows definite logical patterns in complete conformity with the mathematical scheme of quantum theory. The result of these attempts by Birkhoff and Neumann and more recently by Weizsäcker can be stated by saying that the mathematical scheme of quantum theory can be interpreted as an extension or modification of classical logic. It is especially one fundamental principle of classical logic which seems to require a modification. In classical logic it is assumed that, if a statement has any meaning at all, either the statement or the negation of the statement must be correct. Of "here is a table" or "here is not a table," either the first or the second statement must be correct. "Tertium non datur," a third possibility does not exist. It may be that we do not know whether the statement or its negation is correct; but in "reality" one of the two is correct.

In quantum theory this law "tertium non datur" is to be modified. Against any modification of this fundamental principle one can of course at once argue that the principle is assumed in common language and that we have to speak at least about our eventual modification of logic in the natural language. Therefore, it would be a self-contradiction to describe in natural language a logical scheme that does not apply to natural language. There, however, Weizsäcker points out that one may distinguish various levels of language.

One level refers to the objects—for instance, to the atoms or the electrons. A second level refers to statements about objects. A third level may refer to statements about statements about objects, etc. It would then be possible to have different logical patterns at the different levels. It is true that finally we have to go back to the natural language and thereby to the classical logical patterns. But Weizsäcker suggests that classical logic may be in a similar manner a priori to quantum logic, as classical physics is to quantum theory. Classical logic would then be contained as a kind of limiting case in quantum logic, but the latter would constitute the more general logical pattern.

The possible modification of the classical logical pattern shall, then, first refer to the level concerning the objects. Let us consider an atom moving in a closed box which is divided by a wall into two equal parts. The wall may have a very small hole so that the atom can go through. Then the atom can, according to classical logic, be either in the left half of the box or in the right half. There is no third possibility: "tertium non datur." In quantum theory, however, we have to admit—if we use the words "atom" and "box" at all—that there are other possibilities which are in a strange way mixtures of the two former possibilities. This is necessary for explaining the results of our experiments. We could, for instance, observe light that has been scattered by the atom. We could perform three experiments: first the atom is (for instance, by closing the hole in the wall) confined to the left half of the box, and the intensity distribution of the scattered light is measured; then it is confined to the right half and again the scattered light is measured; and finally the atom can move freely in the whole box and again the intensity distribution of the scattered light is measured. If the atom would always be in either the left half or the right half of the box, the final intensity distribution should be a mixture (according to the fraction of time spent by the atom in each of the two parts) of the two former intensity distributions. But this is in general not true experimentally. The real intensity distribution is modified by the "interference of probabilities"; this has been discussed before.

In order to cope with this situation Weizsäcker has introduced the concept "degree of truth." For any simple statement in an alternative like "The atom is in the left (or in the right) half of the box" a complex number is defined as a measure for its "degree of truth." If the number is 1, it means that the statement is true; if the number is 0, it means that it is false. But other values are possible. The absolute square of the complex number gives the probability for the statement's being true; the sum of the two probabilities referring to the two parts in the alternative (either "left" or "right" in our case) must be unity. But each pair of complex numbers referring to the two parts of the alternative represents, according to Weizsäcker's definitions, a "statement" which is certainly true if the numbers have just these values; the two numbers, for instance, are sufficient for determining the intensity distribution of scattered light in our experiment. If one allows the use of the term "statement" in this way one can introduce the term "complementarity" by the following definition: Each statement that is not identical with either of the two alternative statements—in our case with the statements: "the atom is in the left half" or "the atom is in the right half of the box"—is called complementary to these statements. For each complementary statement the question whether the atom is left or right is not decided. But the term "not decided" is by no means equivalent to the term "not known." "Not known" would mean that the atom is "really" left or right, only we do not know where it is. But "not decided" indicates a different situation, expressible only by a complementary statement.

This general logical pattern, the details of which cannot be described here, corresponds precisely to the mathematical formalism of quantum theory. It forms the basis of a precise language that can be used to describe the structure of the atom. But the application of such a language raises a number of difficult problems of which we shall discuss only two here: the relation between the different "levels" of language and the consequences for the underlying ontology.

In classical logic the relation between the different levels of language is a one-to-one correspondence. The two statements, "The atom is in the left half" and "It is true that the atom is in the left half," belong logically to different levels. In classical logic these statements are completely equivalent, i.e., they are either both true or both false. It is not possible that the one is true and the other false. But in the logical pattern of complementarity this relation is more complicated. The correctness or incorrectness of the first statement still implies the correctness or incorrectness of the second statement. But the incorrectness of the second statement does not imply the incorrectness of the first statement. If the second statement is incorrect, it may be undecided whether the atom is in the left half; the atom need not necessarily be in the right half. There is still complete equivalence between the two levels of language with respect to the correctness of a statement, but not with respect to the incorrectness. From this connection one can understand the persistence of the classical laws in quantum theory: wherever a definite result can be derived in a given experiment by the application of the classical laws the result will also follow from quantum theory, and it will hold experimentally.

The final aim of Weizsäcker's attempt is to apply the modified logical patterns also in the higher levels of language, but these questions cannot be discussed here.

The other problem concerns the ontology that underlies the modified logical patterns. If the pair of complex numbers represents a "statement" in the sense just described, there should exist a "state" or a "situation" in nature in which the statement is correct. We will use the word "state" in this connection. The "states" corresponding to complementary statements are then called "coexistent states" by Weizsäcker. This term "coexistent" describes the situation correctly; it would in fact be difficult to call them "different states," since every state contains to some extent also the other "coexistent states." This concept of "state" would then form a first definition concerning the ontology of quantum theory. One sees at once that this use of the word "state," especially the term "coexistent state," is so different from the usual materialistic ontology that one may doubt whether one is using a convenient terminology. On the other hand, if one considers the word "state" as describing some potentiality rather than a reality—one may even simply replace the term "state" by the term "potentiality"—then the concept of "coexistent potentialities" is quite plausible, since one potentiality may involve or overlap other potentialities.

All these difficult definitions and distinctions can be avoided if one confines the language to the description of facts, i.e., experimental results. However, if one wishes to speak about the atomic particles themselves one must either use the mathematical scheme as the only supplement to natural language or one must combine it with a language that makes use of a modified logic or of no well-defined logic at all. In the experiments about atomic events we have to do with things and facts, with phenomena that are just as real as any phenomena in daily life. But the atoms or the elementary particles themselves are not as real; they form a world of potentialities or possibilities rather than one of things or facts.

XI.

The Role of Modern Physics in the Present Development of Human Thinking

THE philosophical implications of modern physics have been discussed in the foregoing chapters in order to show that this most modern part of science touches very old trends of thought at many points, that it approaches some of the very old problems from a new direction. It is probably true quite generally that in the history of human thinking the most fruitful developments frequently take place at those points where two different lines of thought meet. These lines may have their roots in quite different parts of human culture, in different times or different cultural environments or different religious traditions; hence if they actually meet, that is, if they are at least so much related to each other that a real interaction can take place, then one may hope that new and interesting developments will follow. Atomic physics as a part of modern science does actually penetrate in our time into very different cultural traditions. It is not only taught in Europe and the Western countries, where it belongs to the traditional activity in the natural sciences, but it is also studied in the Far East, in countries like Japan and China and India, with their quite different cultural background, and in Russia, where a new way of thinking has been established in our time; a new way related both to specific scientific developments of the Europe of the nineteenth century and to other entirely different traditions from Russia itself. It can certainly not be the purpose of the following discussion to make predictions about the probable result of the encounter between the ideas of modern physics and the older traditions. But it may be possible to define the points from which the interaction between the different ideas may begin.

In considering this process of expansion of modern physics it would certainly not be possible to separate it from the general expansion of natural science, of industry and engineering, of medicine, etc., that is, quite generally of modern civilization in all parts of the world. Modern physics is just one link in a long chain of events that started from the work of Bacon, Galileo and Kepler and from the practical application of natural science in the seventeenth and eighteenth centuries. The connection between natural science and technical science has from the beginning been that of mutual assistance: The progress in technical science, the improvement of the tools, the invention of new technical devices have provided the basis for more, and more accurate, empirical knowledge of nature; and the progress in the understanding of nature and finally the mathematical formulation of natural laws have opened the way to new applications of this knowledge in technical science. For instance, the invention of the telescope enabled the astronomers to measure the motion of the stars more accurately than before; thereby a considerable progress in astronomy and in mechanics was made possible. On the other hand, precise knowledge of the mechanical laws was of the greatest value for the improvement of mechanical tools, for the construction of engines, etc. The great expansion of this combination of natural and technical science started when one had succeeded in putting some of the forces of nature at the disposal of man. The energy stored up in coal, for instance, could then perform some of the work which formerly had to be done by man himself. The industries growing out of these new possibilities could first be considered as a natural continuation and expansion of the older trades; at many points the work of the machines still resembled the older handicraft and the work in the chemical factories could be considered as a continuation of the work in the dyehouses and the pharmacies of the older times. But later entirely new branches of industry developed which had no counterpart in the older trades; for instance, electrical engineering. The penetration of science into the more remote parts of nature enabled the engineers to use forces of nature which in former periods had scarcely been known; and the accurate knowledge of these forces in terms of a mathematical formulation of the laws governing them formed a solid basis for the construction of all kinds of machinery.

The enormous success of this combination of natural and technical science led to a strong preponderance of those nations or states or communities in which this kind of human activity flourished, and as a natural consequence this activity had to be taken up even by those nations which by tradition would not have been inclined toward natural and technical sciences. The modern means of communication and of traffic finally completed this process of expansion of technical civilization. Undoubtedly the process has fundamentally changed the conditions of life on our earth; and whether one approves of it or not, whether one calls it progress or danger, one must realize that it has gone far beyond any control through human forces. One may rather consider it as a biological process on the largest scale whereby the structures active in the human organism encroach on larger parts of matter and transform it into a state suited for the increasing human population.

Modern physics belongs to the most recent parts of this development, and its unfortunately most visible result, the invention of nuclear weapons, has shown the essence of this development in the sharpest possible light. On the one hand, it has demonstrated most clearly that the changes brought about by the combination of natural and technical sciences cannot be looked at only from the optimistic viewpoint; it has at least partly justified the views of those who had always warned against the dangers of such radical transmutation of our natural conditions of life. On the other hand, it has compelled even those nations or individuals who tried to keep apart from these dangers to pay the strongest attention to the new development, since obviously political power in the sense of military power rests upon the possession of atomic weapons. It can certainly not be the task of this volume to discuss extensively the political implications of nuclear physics. But at least a few words may be said about these problems because they always come first into the minds of people when atomic physics is mentioned.

It is obvious that the invention of the new weapons, especially of the thermonuclear weapons, has fundamentally changed the political structure of the world. Not only has the concept of independent nations or states undergone a decisive change, since any nation which is not in possession of such weapons must depend in some way on those very few nations that do produce these arms in large quantity; but also the attempt of warfare on a large scale by means of such weapons has become practically an absurd kind of suicide. Hence one frequently hears the optimistic view that therefore war has become obsolete, that it will not happen again. This view, unfortunately, is a much too optimistic oversimplification. On the contrary, the absurdity of warfare by means of thermonuclear weapons may, in a first approximation, act as an incentive for war on a small scale. Any nation or political group which is convinced of its historical or moral right to enforce some change of the present situation will feel that the use of conventional arms for this purpose will not involve any great risks; they will assume that the other side will certainly not have recourse to the nuclear weapons, since the other side being historically and morally wrong in this issue will not take the chance of war on a large scale. This situation would in turn induce the other nations to state that in case of small wars inflicted upon them by aggressors, they would actually have recourse to the nuclear weapons, and thus the danger obviously remains. It may quite well be that in about twenty or thirty years from now the world will have undergone so great changes that the danger of warfare on a large scale, of the application of all technical resources for the annihilation of the opponent, will have greatly diminished or disappeared. But the way to this new state will be full of the greatest dangers. We must as in all former times, realize that what looks historically or morally right to the one side may look wrong to the other side. The continuation of the status quo may not always be the correct solution; it may, on the contrary, be most important to find peaceful means of adjustments to new situations, and it may in many cases be extremely difficult to find any just decision at all. Therefore, it is probably not too pessimistic to say that the great war can be avoided only if all the different political groups are ready to renounce some of their apparently most obvious rights—in view of the fact that the question of right or wrong may look essentially different from the other side. This is certainly not a new point of view; it is in fact only an application of that human attitude which has been taught through many centuries by some of the great religions.

The invention of nuclear weapons has also raised entirely new problems for science and scientists. The political influence of science has become very much stronger than it was before World War II, and this fact has burdened the scientist, especially the atomic physicist, with a double responsibility. He can either take an active part in the administration of the country in connection with the importance of science for the community; then he will eventually have to face the responsibility for decisions of enormous weight which go far beyond the small circle of research and university work to which he was wont. Or he may voluntarily withdraw from any participation in political decisions; then he will still be responsible for wrong decisions which he could possibly have prevented had he not preferred the quiet life of the scientist. Obviously it is the duty of the scientists to inform their governments in detail about the unprecedented destruction that would follow from a war with thermonuclear weapons. Beyond that, scientists are frequently requested to participate in solemn resolutions in favor of world peace; but considering this latter demand I must confess that I have never been able to see any point in declarations of this kind. Such resolutions may seem a welcome proof of goodwill; but anyone who speaks in favor of peace without stating precisely the conditions of this peace must at once be suspected of speaking only about that kind of peace in which he and his group thrive best—which of course would be completely worthless. Any honest declaration for peace must be an enumeration of the sacrifices one is prepared to make for its preservation. But as a rule the scientists have no authority to make statements of this kind.

At the same time the scientist can do his best to promote international co-operation in his own field. The great importance that many governments attach to research in nuclear physics nowadays and the fact that the level of scientific work is still very different in different countries favors international co-operation in this work. Young scientists of many different countries may gather in research institutions in which a strong activity in the field of modern physics is going on and the common work on difficult scientific problems will foster mutual understanding. In one case, that of the Geneva organization, it has even been possible to reach an agreement between a number of different nations for building a common laboratory and for constructing by a combined effort the expensive experimental equipment for research in nuclear physics. This kind of co-operation will certainly help to establish a common attitude toward the problems of science—common even beyond the purely scientific problems—among the younger generation of scientists. Of course one does not know beforehand what will grow out of the seeds that have been sown in this way when the scientists return into their old environments and again take part in their own cultural traditions. But one can scarcely doubt that the exchange of ideas between young scientists of different countries and between the different generations in every country will help to approach without too much tension that new state of affairs in which a balance is reached between the older traditional forces and the inevitable necessities of modern life. It is especially one feature of science which makes it more than anything else suited for establishing the first strong connection between different cultural traditions. This is the fact that the ultimate decisions about the value of a special scientific work, about what is correct or wrong in the work, do not depend on any human authority. It may sometimes take many years before one knows the solution of a problem, before one can distinguish between truth and error; but finally the questions will be decided, and the decisions are made not by any group of scientists but by nature itself. Therefore, scientific ideas spread among those who are interested in science in an entirely different way from the propagation of political ideas.

While political ideas may gain a convincing influence among great masses of people just because they correspond or seem to correspond to the prevailing interests of the people, scientific ideas will spread only because they are true. There are objective and final criteria assuring the correctness of a scientific statement.

All that has here been said about international co-operation and exchange of ideas would of course be equally true for any part of modern science; it is by no means confined to atomic physics. In this respect modern physics is just one of the many branches of science, and even if its technical applications—the arms and the peaceful use of atomic energy—attach a special weight to this branch, there would be no reason for considering international co-operation in this field as far more important than in any other field. But we have now to discuss again those features of modern physics which are essentially different from the previous development of natural science, and we have for this purpose once more to go back to the European history of this development that was brought about by the combination of natural and technical sciences.

It has frequently been discussed among the historians whether the rise of natural science after the sixteenth century was in any way a natural consequence of earlier trends in human thinking.

It may be argued that certain trends in Christian philosophy led to a very abstract concept of God, that they put God so far above the world that one began to consider the world without at the same time also seeing God in the world. The Cartesian partition may be called a final step in this development. Or one may point out that all the theological controversies of the sixteenth century produced a general discontent about problems that could not really be settled by reason and were exposed to the political struggles of the time; that this discontent favored interest in problems which were entirely separated from the theological disputes. Or one may simply refer to the enormous activity, the new spirit that had come into the European societies through the Renaissance. In any case during this period a new authority appeared which was completely independent of Christian religion or philosophy or of the Church, the authority of experience, of the empirical fact. One may trace this authority back into older philosophical trends, for instance, into the philosophy of Occam and Duns Scotus, but it became a vital force of human activity only from the sixteenth century onward. Galileo did not only *think* about the mechanical motions, the pendulum and the falling stone; he tried out by experiments, quantitatively, how these motions took place. This new activity was in its beginning certainly not meant as a deviation from the traditional Christian religion. On the contrary, one spoke of two kinds of revelation of God. The one was written in the Bible and the other was to be found in the book of nature. The Holy Scripture had been written by man and was therefore subject to error, while nature was the immediate expression of God's intentions.

However, the emphasis on experience was connected with a slow and gradual change in the aspect of reality. While in the Middle Ages what we nowadays call the symbolic meaning of a thing was in some way its primary reality, the aspect of reality changed toward what we can perceive with our senses. What we can see and touch became primarily real. And this new concept of reality could be connected with a new activity: we can experiment and see how things really are. It was easily seen that this new attitude meant the departure of the human mind into an immense field of new possibilities, and it can be well understood that the Church saw in the new movement the dangers rather than the hopes. The famous trial of Galileo in connection with his views on the Copernican system marked the beginning of a struggle that went on for more than a century. In this controversy the representatives of natural science could argue that experience offers an undisputable truth, that it cannot be left to any human authority to decide about what really happens in nature, and that this decision is made by nature or in this sense by God. The representatives of the traditional religion, on the other hand, could argue that by paying too much attention to the material world, to what we perceive with our senses, we lose the connection with the essential values of human life, with just that part of reality which is beyond the material world. These two arguments do not meet, and therefore the problem could not be settled by any kind of agreement or decision.

In the meantime natural science proceeded to get a clearer and wider picture of the material world. In physics this picture was to be described by means of those concepts which we nowadays call the concepts of classical physics. The world consisted of things in space and time, the things consist of matter, and matter can produce and can be acted upon by forces. The events follow from the interplay between matter and forces; every event is the result and the cause of other events. At the same time the human attitude toward nature changed from a contemplative one to the pragmatic one. One was not so much interested in nature as it is; one rather asked what one could do with it. Therefore, natural science turned into technical science; every advancement of knowledge was connected with the question as to what practical use could be derived from it. This was true not only in physics; in chemistry and biology the attitude was essentially the same, and the success of the new methods in medicine or in agriculture contributed essentially to the propagation of the new tendencies.

In this way, finally, the nineteenth century developed an extremely rigid frame for natural science which formed not only science but also the general outlook of great masses of people. This frame was supported by the fundamental concepts of classical physics, space, time, matter and causality; the concept of reality applied to the things or events that we could perceive by our senses or that could be observed by means of the refined tools that technical science had provided. Matter was the primary reality. The progress of science was pictured as a crusade of conquest into the material world. Utility was the watchword of the time.

On the other hand, this frame was so narrow and rigid that it was difficult to find a place in it for many concepts of our language that had always belonged to its very substance, for instance, the concepts of mind, of the human soul or of life. Mind could be introduced into the general picture only as a kind of mirror of the material world; and when one studied the properties of this mirror in the science of psychology, the scientists were always tempted—if I may carry the comparison further—to pay more attention to its mechanical than to its optical properties. Even there one tried to apply the concepts of classical physics, primarily that of causality. In the same way life was to be explained as a physical and chemical process, governed by natural laws, completely determined by causality. Darwin's concept of evolution provided ample evidence for this interpretation. It was especially difficult to find in this framework room for those parts of reality that had been the object of the traditional religion and seemed now more or less only imaginary. Therefore, in those European countries in which one was wont to follow the ideas up to their extreme consequences, an open hostility of science toward religion developed, and even in the other countries there was an increasing tendency toward indifference toward such questions; only the ethical values of the Christian religion were excepted from this trend, at least for the time being. Confidence in the scientific method and in rational thinking replaced all other safeguards of the human mind.

Coming back now to the contributions of modern physics, one may say that the most important change brought about by its results consists in the dissolution of this rigid frame of concepts of the nineteenth century. Of course many attempts had been made before to get away from this rigid frame which seemed obviously too narrow for an understanding of the essential parts of reality. But it had not been possible to see what could be wrong with the fundamental concepts like matter, space, time and causality that had been so extremely successful in the history of science. Only experimental research itself, carried out with all the refined equipment that technical science could offer, and its mathematical interpretation, provided the basis for a critical analysis—or, one may say, enforced the critical analysis—of these concepts, and finally resulted in the dissolution of the rigid frame.

This dissolution took place in two distinct stages. The first was the discovery, through the theory of relativity, that even such fundamental concepts as space and time could be changed and in fact must be changed on account of new experience. This change did not concern the somewhat vague concepts of space and time in natural language; but it did concern their precise formulation in the scientific language of Newtonian mechanics, which had erroneously been accepted as final. The second stage was the discussion of the concept of matter enforced by the experimental results concerning the atomic structure. The idea of the reality of matter had probably been the strongest part in that rigid frame of concepts of the nineteenth century, and this idea had at least to be modified in connection with the new experience. Again the concepts so far as they belonged to the natural language remained untouched. There was no difficulty in speaking about matter or about facts or about reality when one had to describe the atomic experiments and their results. But the scientific extrapolation of these concepts into the smallest parts of matter could not be done in the simple way suggested by classical physics, though it had erroneously determined the general outlook on the problem of matter.

These new results had first of all to be considered as a serious warning against the somewhat forced application of scientific concepts in domains where they did not belong. The application of the concepts of classical physics, e.g., in chemistry, had been a mistake. Therefore, one will nowadays be less inclined to assume that the concepts of physics, even those of quantum theory, can certainly be applied everywhere in biology or other sciences. We will, on the contrary, try to keep the doors open for the entrance of new concepts even in those parts of science where the older concepts have been very useful for the understanding of the phenomena. Especially at those points where the application of the older concepts seems somewhat forced or appears not quite adequate to the problem we will try to avoid any rash conclusions.

Furthermore, one of the most important features of the development and the analysis of modern physics is the experience that the concepts of natural language, vaguely defined as they are, seem to be more stable in the expansion of knowledge than the precise terms of scientific language, derived as an idealization from only limited groups of phenomena. This is in fact not surprising since the concepts of natural language are formed by the immediate connection with reality; they represent reality. It is true that they are not very well defined and may therefore also undergo changes in the course of the centuries, just as reality itself did, but they never lose the immediate connection with reality. On the other hand, the scientific concepts are idealizations; they are derived from experience obtained by refined experimental tools, and are precisely defined through axioms and definitions. Only through these precise definitions is it possible to connect the concepts with a mathematical scheme and to derive mathematically the infinite variety of possible phenomena in this field. But through this process of idealization and precise definition the immediate connection with reality is lost. The concepts still correspond very closely to reality in that part of nature which had been the object of the research. But the correspondence may be lost in other parts containing other groups of phenomena.

Keeping in mind the intrinsic stability of the concepts of natural language in the process of scientific development, one sees that—after the experience of modern physics—our attitude toward concepts like mind or the human soul or life or God will be different from that of the nineteenth century, because these concepts belong to the natural language and have therefore immediate connection with reality. It is true that we will also realize that these concepts are not well defined in the scientific sense and that their application may lead to various contradictions, for the time being we may have to take the concepts, unanalyzed as they are; but still we know that they touch reality. It may be useful in this connection to remember that even in the most precise part of science, in mathematics, we cannot avoid using concepts that involve contradictions. For instance, it is well known that the concept of infinity leads to contradictions that have been analyzed, but it would be practically impossible to construct the main parts of mathematics without this concept.

The general trend of human thinking in the nineteenth century had been toward an increasing confidence in the scientific method and in precise rational terms, and had led to a general skepticism with regard to those concepts of natural language which do not fit into the closed frame of scientific thought—for instance, those of religion. Modern physics has in many ways increased this skepticism; but it has at the same time turned it against the overestimation of precise scientific concepts, against a too-optimistic view on progress in general, and finally against skepticism itself. The skepticism against precise scientific concepts does not mean that there should be a definite limitation for the application of rational thinking. On the contrary, one may say that the human ability to understand may be in a certain sense unlimited. But the existing scientific concepts cover always only a very limited part of reality, and the other part that has not yet been understood is infinite. Whenever we proceed from the known into the unknown we may hope to understand, but we may have to learn at the same time a new meaning of the word "understanding." We know that any understanding must be based finally upon the natural language because it is only there that we can be certain to touch reality, and hence we must be skeptical about any skepticism with regard to this natural language and its essential concepts. Therefore, we may use these concepts as they have been used at all times. In this way modern physics has perhaps opened the door to a wider outlook on the relation between the human mind and reality.

This modern science, then, penetrates in our time into other parts of the world where the cultural tradition has been entirely different from the European civilization. There the impact of this new activity in natural and technical science must make itself felt even more strongly than in Europe, since changes in the conditions of life that have taken two or three centuries in Europe will take place there within a few decades. One should expect that in many places this new activity must appear as a decline of the older culture, as a ruthless and barbarian attitude, that upsets the sensitive balance on which all human happiness rests. Such consequences cannot be avoided; they must be taken as one aspect of our time. But even there the openness of modern physics may help to some extent to reconcile the older traditions with the new trends of thought. For instance, the great scientific contribution in theoretical physics that has come from Japan since the last war may be an indication for a certain relationship between philosophical ideas in the tradition of the Far East and the philosophical substance of quantum theory. It may be easier to adapt oneself to the quantum-theoretical concept of reality when one has not gone through the naïve materialistic way of thinking that still prevailed in Europe in the first decades of this century.

Of course such remarks should not be misunderstood as an underestimation of the damage that may be done or has been done to old cultural traditions by the impact of technical progress. But since this whole development has for a long time passed far beyond any control by human forces, we have to accept it as one of the most essential features of our time and must try to connect it as much as possible with the human values that have been the aim of the older cultural and religious traditions. It may be allowed at this point to quote a story from the Hasidic religion: There was an old rabbi, a priest famous for his wisdom, to whom all people came for advice. A man visited him in despair over all the changes that went on around him, deploring all the harm done by so-called technical progress. "Isn't all this technical nuisance completely worthless," he exclaimed "if one considers the real values of life?" "This may be so," the rabbi replied, "but if one has the right attitude one can learn from everything." "No," the visitor rejoined, "from such foolish things as railway or telephone or telegraph one can learn nothing whatsoever." But the rabbi answered, "You are wrong. From the railway you can learn that you may by being one instant late miss everything. From the telegraph you can learn that every word counts. And from the telephone you can learn that what we say here can be heard there." The visitor understood what the rabbi meant and went away.

Finally, modern science penetrates into those large areas of our present world in which new doctrines were established only a few decades ago as foundations for new and powerful societies. There modern science is confronted both with the content of the doctrines, which go back to European philosophical ideas of the nineteenth century (Hegel and Marx), and with the phenomenon of uncompromising belief. Since modern physics must play a great role in these countries because of its practical applicability, it can scarcely be avoided that the narrowness of the doctrines is felt by those who have really understood modern physics and its philosophical meaning. Therefore, at this point an interaction between science and the general trend of thought may take place. Of course the influence of science should not be overrated; but it might be that the openness of modern science could make it easier even for larger groups of people to see that the doctrines are possibly not so important for the society as had been assumed before. In this way the influence of modern science may favor an attitude of tolerance and thereby may prove valuable.

On the other hand, the phenomenon of uncompromising belief carries much more weight than some special philosophical notions of the nineteenth century. We cannot close our eyes to the fact that the great majority of the people can scarcely have any well-founded judgment concerning the correctness of certain important general ideas or doctrines. Therefore, the word "belief" can for this majority not mean "perceiving the truth of something" but can only be understood as "taking this as the basis for life." One can easily understand that this second kind of belief is much firmer, is much more fixed than the first one, that it can persist even against immediate contradicting experience and can therefore not be shaken by added scientific knowledge. The history of the past two decades has shown by many examples that this second kind of belief can sometimes be upheld to a point where it seems completely absurd, and that it then ends only with the death of the believer. Science and history can teach us that this kind of belief may become a great danger for those who share it. But such knowledge is of no avail, since one cannot see how it could be avoided, and therefore such belief has always belonged to the great forces in human history. From the scientific tradition of the nineteenth century one would of course be inclined to hope that all belief should be based on a rational analysis of every argument, on careful deliberation; and that this other kind of belief, in which some real or apparent truth is simply taken as the basis for life, should not exist. It is true that cautious deliberation based on purely rational arguments can save us from many errors and dangers, since it allows readjustment to new situations, and this may be a necessary condition for life. But remembering our experience in modern physics it is easy to see that there must always be a fundamental complementarity between deliberation and decision. In the practical decisions of life it will scarcely ever be possible to go through all the arguments in favor of or against one possible decision, and one will therefore always have to act on insufficient evidence. The decision finally takes place by pushing away all the arguments—both those that have been understood and others that might come up through further deliberation—and by cutting off all further pondering. The decision may be the result of deliberation, but it is at the same time complementary to deliberation; it excludes deliberation. Even the most important decisions in life must always contain this inevitable element of irrationality. The decision itself is necessary, since there must be something to rely upon, some principle to guide our actions. Without such a firm stand our own actions would lose all force. Therefore, it cannot be avoided that some real or apparent truth form the basis of life; and this fact should be acknowledged with regard to those groups of people whose basis is different from our own.

Coming now to a conclusion from all that has been said about modern science, one may perhaps state that modern physics is just one, but a very characteristic, part of a general historical process that tends toward a unification and a widening of our present world. This process would in itself lead to a diminution of those cultural and political tensions that create the great danger of our time. But it is accompanied by another process which acts in the opposite direction. The fact that great masses of people become conscious of this process of unification leads to an instigation of all forces in the existing cultural communities that try to ensure for their traditional values the largest possible role in the final state of unification. Thereby the tensions increase and the two competing processes are so closely linked with each other that every intensification of the unifying process—for instance, by means of new technical progress—intensifies also the struggle for influence in the final state, and thereby adds to the instability of the transient state. Modern physics plays perhaps only a small role in this dangerous process of unification. But it helps at two very decisive points to guide the development into a calmer kind of evolution. First, it shows that the use of arms in the process would be disastrous and, second, through its openness for all kinds of concepts it raises the hope that in the final state of unification many different cultural traditions may live together and may combine different human endeavors into a new kind of balance between thought and deed, between activity and meditation.

EPILOGUE

By Ruth Nanda Anshen

EPILOGUE

What World Perspectives Means

by Ruth Nanda Anshen

This is a reprint of Volume XIX of the WORLD PERSPECTIVES SERIES, which the present writer has planned and edited in collaboration with a Board of Editors consisting of NIELS BOHR, RICHARD COURANT, HU SHIH, ERNEST JACKH, ROBERT M. MACIVER, JACQUES MARITAIN, J. ROBERT OPPENHEIMER, I. I. RABI, SARVEPALLI RADHAKRISHNAN, ALEXANDER SACHS.

This volume is part of a plan to present short books in a variety of fields by the most responsible of contemporary thinkers. The purpose is to reveal basic new trends in modern civilization, to interpret the creative forces at work in the East as well as in the West, and to point to the new consciousness which can contribute to a deeper understanding of the interrelation of man and the universe, the individual and society, and of the values shared by all people. *World Perspectives* represents the world community of ideas in a universe of discourse, emphasizing the principle of unity in mankind of permanence within change.

Recent developments in many fields of thought have opened unsuspected prospects for a deeper understanding of man's situation and for a proper appreciation of human values and human aspirations. These prospects, though the outcome of purely specialized studies in limited fields, require for their analysis and synthesis a new structure and frame in which they can be explored, enriched and advanced in all their aspects for the benefit of man and society. Such a structure and frame it is the endeavor of *World Perspectives* to define, leading hopefully to a doctrine of man.

A further purpose of this Series is to attempt to overcome a principal ailment of humanity, namely, the effects of the atomization of knowledge produced by the overwhelming accretion of facts which science has created; to clarify and synthesize ideas through the *depth* fertilization of minds; to show from diverse and important points of view the correlation of ideas, facts and values which are in perpetual interplay; to demonstrate the character, kinship, logic and operation of the entire organism of reality while showing the persistent interrelationship of the processes of the human mind and in the interstices of knowledge; to reveal the inner synthesis and organic unity of life itself.

It is the thesis of *World Perspectives* that in spite of the difference and diversity of the disciplines represented, there exists a strong common agreement among the authors concerning the overwhelming need for counterbalancing the multitude of compelling scientific activities and investigations of objective phenomena from physics to metaphysics, history and biology and to relate these to meaningful experience. To provide this balance, it is necessary to stimulate an awareness of the basic fact that ultimately the individual human personality must tie all the loose ends together into an organic whole, must relate himself to himself, to mankind and society while deepening and enhancing his communion with the universe. To anchor this spirit and to impress it on the intellectual and spiritual life of humanity, on thinkers and doers alike, is indeed an enormous challenge which cannot be left entirely either to natural science on the one hand nor to organized religion on the other. For we are confronted with the unbending necessity to discover a principle of differentiation yet relatedness lucid enough to justify and purify scientific, philosophic and all other knowledge while accepting their mutual interdependence. This is the crisis in consciousness made articulate through the crisis in science. This is the new awakening.

World Perspectives is dedicated to the task of showing that basic theoretical knowledge is related to the dynamic content of the wholeness of life. It is dedicated to the new synthesis at once cognitive and intuitive. It is concerned with the unity and continuity of knowledge in relation to man's nature and his understanding, a task for the synthetic imagination and its unifying vistas. Man's situation is new and his response must be new. For the nature of man is knowable in many different ways and all of these paths of knowledge are interconnectable and some are interconnected, like a great network, a great network of people, between ideas, between systems of knowledge, a rationalized kind of structure which is human culture and human society.

Knowledge, it is shown in these volumes, no longer consists in a manipulation of man and nature as opposite forces, nor in the reduction of data to statistical order, but is a means of liberating mankind from the destructive power of fear, pointing the way toward the goal of the rehabilitation of the human will and the rebirth of faith and confidence in the human person. The works published also endeavor to reveal that the cry for patterns, systems and authorities is growing less insistent as the desire grows stronger in both East and West for the recovery of a dignity, integrity and self-realization which are the inalienable rights of man who is not a mere *tabula rasa* on which anything may be arbitrarily imprinted by external circumstance but who possesses the unique potentiality of free creativity. Man is differentiated from other forms of life in that he may guide change by means of conscious purpose in the light of rational experience.

World Perspectives is planned to gain insight into the meaning of man who not only is determined by history but who also determines history. History is to be understood as concerned not only with the life of man on this planet but as including also such cosmic influences as interpenetrate our human world. This generation is discovering that history does not conform to the social optimism of modern civilization and that the organization of human communities and the establishment of freedom, justice and peace are not only intellectual achievements but spiritual and moral achievements as well, demanding a cherishing of the wholeness of human personality, the "unmediated wholeness of feeling and thought," and constituting a never-ending challenge to man, emerging from the abyss of meaninglessness and suffering, to be renewed and replenished in the totality of his life.

World Perspectives is committed to the recognition that all great changes are preceded by a vigorous intellectual reevaluation and reorganization. Our authors are aware that the sin of hybris may be avoided by showing that the creative process itself is not a free activity if by free we mean arbitrary or unrelated to cosmic law. For the creative process in the human mind, the developmental process in organic nature and the basic laws of the inorganic realm may be but varied expressions of a universal formative process. Thus *World Perspectives* hopes to show that although the present apocalyptic period is one of exceptional tensions, there is also an exceptional movement at work toward a compensating unity which cannot obliterate the ultimate moral power pervading the universe, that very power on which all human effort must at last depend. In this way, we may come to understand that there exists an independence of spiritual and mental growth which though conditioned by circumstances is never determined by circumstances. In this way the great plethora of human knowledge may be correlated with an insight into the nature of human nature by being attuned to the wide and deep range of human thought and human experience. For what is lacking is not the knowledge of the structure of the universe but a consciousness of the qualitative uniqueness of human life.

And finally, it is the thesis of this Series that man is in the process of developing a new awareness which, in spite of his apparent spiritual and moral captivity, can eventually lift the human race above and beyond the fear, ignorance, brutality and isolation which beset it today. It is to this nascent consciousness, to this concept of man born out of a fresh vision of reality, that *World Perspectives* is dedicated.

Revised January, 1970

harper torchbooks

American Studies: General

HENRY STEELE COMMAGER, Ed.: The Struggle for Racial Equality TB/1300

CARL N. DEGLER: Out of Our Past: *The Forces that Shaped Modern America* CN/2

CARL N. DEGLER, Ed.: Pivotal Interpretations of American History Vol. I TB/1240; Vol. II TB/1241

A. S. EISENSTADT, Ed.: The Craft of American History: *Selected Essays* Vol. I TB/1255; Vol. II TB/1256

ROBERT L. HEILBRONER: The Limits of American Capitalism TB/1305

JOHN HIGHAM, Ed.: The Reconstruction of American History TB/1068

ROBERT H. JACKSON: The Supreme Court in the American System of Government TB/1106

JOHN F. KENNEDY: A Nation of Immigrants. *Illus. Revised and Enlarged. Introduction by Robert F. Kennedy* TB/1118

RICHARD B. MORRIS: Fair Trial: *Fourteen Who Stood Accused, from Anne Hutchinson to Alger Hiss* TB/1335

GUNNAR MYRDAL: An American Dilemma: *The Negro Problem and Modern Democracy. Introduction by the Author.* Vol. I TB/1443; Vol. II TB/1444

GILBERT OSOFSKY, Ed.: The Burden of Race: *A Documentary History of Negro-White Relations in America* TB/1405

ARNOLD ROSE: The Negro in America: *The Condensed Version of Gunnar Myrdal's* An American Dilemma. *Second Edition* TB/3048

JOHN E. SMITH: Themes in American Philosophy: *Purpose, Experience and Community* TB/1466

WILLIAM R. TAYLOR: Cavalier and Yankee: *The Old South and American National Character* TB/1474

American Studies: Colonial

BERNARD BAILYN: The New England Merchants in the Seventeenth Century TB/1149

ROBERT E. BROWN: Middle-Class Democracy and Revolution in Massachusetts, 1691–1780. *New Introduction by Author* TB/1413

JOSEPH CHARLES: The Origins of the American Party System TB/1049

WESLEY FRANK CRAVEN: The Colonies in Transition: 1660-1712† TB/3084

CHARLES GIBSON: Spain in America † TB/3077

CHARLES GIBSON, Ed.: The Spanish Tradition in America + HR/1351

LAWRENCE HENRY GIPSON: The Coming of the Revolution: 1763-1775. † *Illus.* TB/3007

PERRY MILLER: Errand Into the Wilderness TB/1139

PERRY MILLER & T. H. JOHNSON, Eds.: The Puritans: *A Sourcebook of Their Writings* Vol. I TB/1093; Vol. II TB/1094

EDMUND S. MORGAN: The Puritan Family: *Religion and Domestic Relations in Seventeenth Century New England* TB/1227

WALLACE NOTESTEIN: The English People on the Eve of Colonization: 1603-1630. † *Illus.* TB/3006

LOUIS B. WRIGHT: The Cultural Life of the American Colonies: 1607-1763. † *Illus.* TB/3005

American Studies: The Revolution to 1860

JOHN R. ALDEN: The American Revolution: 1775-1783. † *Illus.* TB/3011

RAY A. BILLINGTON: The Far Western Frontier: 1830-1860. † *Illus.* TB/3012

GEORGE DANGERFIELD: The Awakening of American Nationalism, 1815-1828. † *Illus.* TB/3061

CLEMENT EATON: The Growth of Southern Civilization, 1790-1860. † *Illus.* TB/3040

LOUIS FILLER: The Crusade against Slavery: 1830-1860. † *Illus.* TB/3029

WILLIM W. FREEHLING: Prelude to Civil War: *The Nullification Controversy in South Carolina, 1816-1836* TB/1359

THOMAS JEFFERSON: Notes on the State of Virginia. ‡ *Edited by Thomas P. Abernethy* TB/3052

JOHN C. MILLER: The Federalist Era: 1789-1801. † *Illus.* TB/3027

RICHARD B. MORRIS: The American Revolution Reconsidered TB/1363

GILBERT OSOFSKY, Ed.: Puttin' On Ole Massa: *The Slave Narratives of Henry Bibb, William Wells Brown, and Solomon Northup* ‡ TB/1432

FRANCIS S. PHILBRICK: The Rise of the West, 1754-1830. † *Illus.* TB/3067

MARSHALL SMELSER: The Democratic Republic, 1801-1815 † TB/1406

† The New American Nation Series, edited by Henry Steele Commager and Richard B. Morris.
‡ American Perspectives series, edited by Bernard Wishy and William E. Leuchtenburg.
α History of Europe series, edited by J. H. Plumb.
§ The Library of Religion and Culture, edited by Benjamin Nelson.
‖ Researches in the Social, Cultural, and Behavioral Sciences, edited by Benjamin Nelson.
Σ Harper Modern Science Series, edited by James A. Newman.
° Not for sale in Canada.
+ Documentary History of the United States series, edited by Richard B. Morris.
# Documentary History of Western Civilization series, edited by Eugene C. Black and Leonard W. Levy.
^ The Economic History of the United States series, edited by Henry David et al.
¶ European Perspectives series, edited by Eugene C. Black.
** Contemporary Essays series, edited by Leonard W. Levy.
* The Stratum Series, edited by John Hale.

LOUIS B. WRIGHT: Culture on the Moving Frontier TB/1053

American Studies: The Civil War to 1900

T. C. COCHRAN & WILLIAM MILLER: The Age of Enterprise: *A Social History of Industrial America* TB/1054
W. A. DUNNING: Reconstruction, Political and Economic: 1865-1877 TB/1073
HAROLD U. FAULKNER: Politics, Reform and Expansion: 1890-1900. † *Illus.* TB/3020
GEORGE M. FREDRICKSON: The Inner Civil War: *Northern Intellectuals and the Crisis of the Union* TB/1358
JOHN A. GARRATY: The New Commonwealth, 1877-1890 † TB/1410
HELEN HUNT JACKSON: A Century of Dishonor: *The Early Crusade for Indian Reform. † Edited by Andrew F. Rolle* TB/3063
WILLIAM G. MCLOUGHLIN, Ed.: The American Evangelicals, 1800-1900: An Anthology ‡ TB/1382
JAMES S. PIKE: The Prostrate State: *South Carolina under Negro Government. ‡ Intro. by Robert F. Durden* TB/3085
VERNON LANE WHARTON: The Negro in Mississippi, 1865-1890 TB/1178

American Studies: The Twentieth Century

RAY STANNARD BAKER: Following the Color Line: *American Negro Citizenship in Progressive Era. ‡ Edited by Dewey W. Grantham, Jr. Illus.* TB/3053
RANDOLPH S. BOURNE: War and the Intellectuals: *Collected Essays, 1915-1919. ‡ Edited by Carl Resek* TB/3043
A. RUSSELL BUCHANAN: The United States and World War II. † *Illus.* Vol. I TB/3044; Vol. II TB/3045
THOMAS C. COCHRAN: The American Business System: *A Historical Perspective, 1900-1955* TB/1080
FOSTER RHEA DULLES: America's Rise to World Power: 1898-1954. † *Illus.* TB/3021
HAROLD U. FAULKNER: The Decline of Laissez Faire, 1897-1917 TB/1397
JOHN D. HICKS: Republican Ascendancy: 1921-1933. † *Illus.* TB/3041
WILLIAM E. LEUCHTENBURG: Franklin D. Roosevelt and the New Deal: 1932-1940. † *Illus.* TB/3025
WILLIAM E. LEUCHTENBURG, Ed.: The New Deal: *A Documentary History* + HR/1354
ARTHUR S. LINK: Woodrow Wilson and the Progressive Era: 1910-1917. † *Illus.* TB/3023
BROADUS MITCHELL: Depression Decade: *From New Era through New Deal, 1929-1941* ^ TB/1439
GEORGE E. MOWRY: The Era of Theodore Roosevelt and the Birth of Modern America: 1900-1912. † *Illus.* TB/3022
WILLIAM PRESTON, JR.: Aliens and Dissenters:
TWELVE SOUTHERNERS: I'll Take My Stand: *The South and the Agrarian Tradition. Intro. by Louis D. Rubin, Jr.; Biographical Essays by Virginia Rock* TB/1072

Art, Art History, Aesthetics

ERWIN PANOFSKY: Renaissance and Renascences in Western Art. *Illus.* TB/1447
ERWIN PANOFSKY: Studies in Iconology: *Humanistic Themes in the Art of the Renaissance. 180 illus.* TB/1077
HEINRICH ZIMMER: Myths and Symbols in Indian Art and Civilization. *70 illus.* TB/2005

Asian Studies

WOLFGANG FRANKE: China and the West: *Th Cultural Encounter, 13th to 20th Centurie Trans. by R. A. Wilson* TB/132
L. CARRINGTON GOODRICH: A Short History (the Chinese People. *Illus.* TB/301

Economics & Economic History

C. E. BLACK: The Dynamics of Modernizatior *A Study in Comparative History* TB/132
GILBERT BURCK & EDITORS OF *Fortune:* The Con puter Age: *And its Potential for Manag ment* TB/117
ROBERT L. HEILBRONER: The Future as History *The Historic Currents of Our Time and th Direction in Which They Are Taking Americ* TB/138
ROBERT L. HEILBRONER: The Great Ascent: *Th Struggle for Economic Development in O Time* TB/303
FRANK H. KNIGHT: The Economic Organizatio TB/121
DAVID S. LANDES: Bankers and Pashas: *Intern tional Finance and Economic Imperialism Egypt. New Preface by the Author* TB/141
ROBERT LATOUCHE: The Birth of Western Ecor omy: *Economic Aspects of the Dark Age* TB/129
W. ARTHUR LEWIS: The Principles of Economi Planning. *New Introduction by the Author* TB/143
WILLIAM MILLER, Ed.: Men in Business: *Essay on the Historical Role of the Entreprene* TB/108
HERBERT A. SIMON: The Shape of Automation *For Men and Management* TB/124

Historiography and History of Ideas

J. BRONOWSKI & BRUCE MAZLISH: The Wester Intellectual Tradition: *From Leonardo t Hegel* TB/300
WILHELM DILTHEY: Pattern and Meaning in Hi tory: *Thoughts on History and Society. Edited with an Intro. by H. P. Rickman* TB/107
J. H. HEXTER: More's Utopia: *The Biography o an Idea. Epilogue by the Author* TB/119
H. STUART HUGHES: History as Art and a Science: *Twin Vistas on the Past* TB/120
ARTHUR O. LOVEJOY: The Great Chain of Being *A Study of the History of an Idea* TB/100
RICHARD H. POPKIN: The History of Scepticisn from Erasmus to Descartes. *Revised Editio* TB/139
BRUNO SNELL: The Discovery of the Mind: *Th Greek Origins of European Thought* TB/101

History: General

HANS KOHN: The Age of Nationalism: *Th First Era of Global History* TB/138
BERNARD LEWIS: The Arabs in History TB/102
BERNARD LEWIS: The Middle East and th West ° TB/127

History: Ancient

A. ANDREWS: The Greek Tyrants TB/110
THEODOR H. GASTER: Thespis: *Ritual Myth an Drama in the Ancient Near East* TB/128

. H. M. JONES, Ed.: A History of Rome through the Fifth Century # *Vol. I: The Republic* HR/1364
Vol. II The Empire: HR/1460

AMUEL NOAH KRAMER: Sumerian Mythology TB/1055

'APHTALI LEWIS & MEYER REINHOLD, Eds.: Roman Civilization *Vol. I: The Republic* TB/1231
Vol. II: The Empire TB/1232

istory: Medieval

ORMAN COHN: The Pursuit of the Millennium: *Revolutionary Messianism in Medieval and Reformation Europe* TB/1037

. L. GANSHOF: Feudalism TB/1058

. L. GANSHOF: The Middle Ages: *A History of International Relations. Translated by Rêmy Hall* TB/1411

ENRY CHARLES LEA: The Inquisition of the Middle Ages. || *Introduction by Walter Ullmann* TB/1456

istory: Renaissance & Reformation

ACOB BURCKHARDT: The Civilization of the Renaissance in Italy. *Introduction by Benjamin Nelson and Charles Trinkaus. Illus.* Vol. I TB/40; Vol. II TB/41

OHN CALVIN & JACOPO SADOLETO: A Reformation Debate. *Edited by John C. Olin* TB/1239

. H. ELLIOTT: Europe Divided, 1559-1598 α ° TB/1414

. R. ELTON: Reformation Europe, 1517-1559 ° α TB/1270

ANS J. HILLERBRAND, Ed., The Protestant Reformation # HR/1342

OHAN HUIZINGA: Erasmus and the Age of Reformation. *Illus.* TB/19

OEL HURSTFIELD: The Elizabethan Nation TB/1312

OEL HURSTFIELD, Ed.: The Reformation Crisis TB/1267

AUL OSKAR KRISTELLER: Renaissance Thought: *The Classic, Scholastic, and Humanist Strains* TB/1048

AVID LITTLE: Religion, Order and Law: *A Study in Pre-Revolutionary England. § Preface by R. Bellah* TB/1418

AOLO ROSSI: Philosophy, Technology, and the Arts, in the Early Modern Era 1400-1700. || *Edited by Benjamin Nelson. Translated by Salvator Attanasio* TB/1458

. R. TREVOR-ROPER: The European Witch-craze of the Sixteenth and Seventeenth Centuries and Other Essays ° TB/1416

istory: Modern European

LAN BULLOCK: Hitler, A Study in Tyranny. ° *Revised Edition. Illus.* TB/1123

OHANN GOTTLIEB FICHTE: Addresses to the German Nation. *Ed. with Intro. by George A. Kelly* ¶ TB/1366

LBERT GOODWIN: The French Revolution TB/1064

TANLEY HOFFMANN et al.: In Search of France: *The Economy, Society and Political System In the Twentieth Century* TB/1219

. STUART HUGHES: The Obstructed Path: *French Social Thought in the Years of Desperation* TB/1451

OHAN HUIZINGA: Dutch Civilisation in the 17th Century and Other Essays TB/1453

JOHN MCMANNERS: European History, 1789-1914: *Men, Machines and Freedom* TB/1419

HUGH SETON-WATSON: Eastern Europe Between the Wars, 1918-1941 TB/1330

ALBERT SOREL: Europe Under the Old Regime. *Translated by Francis H. Herrick* TB/1121

A. J. P. TAYLOR: From Napoleon to Lenin: *Historical Essays* ° TB/1268

A. J. P. TAYLOR: The Habsburg Monarchy, 1809-1918: *A History of the Austrian Empire and Austria-Hungary* ° TB/1187

J. M. THOMPSON: European History, 1494-1789 TB/1431

H. R. TREVOR-ROPER: Historical Essays TB/1269

Literature & Literary Criticism

W. J. BATE: From Classic to Romantic: *Premises of Taste in Eighteenth Century England* TB/1036

VAN WYCK BROOKS: Van Wyck Brooks: The Early Years: *A Selection from his Works, 1908-1921 Ed. with Intro. by Claire Sprague* TB/3082

RICHMOND LATTIMORE, Translator: The Odyssey of Homer TB/1389

ROBERT PREYER, Ed.: Victorian Literature ** TB/1302

Philosophy

HENRI BERGSON: Time and Free Will: *An Essay on the Immediate Data of Consciousness* ° TB/1021

H. J. BLACKHAM: Six Existentialist Thinkers: *Kierkegaard, Nietzsche, Jaspers, Marcel, Heidegger, Sartre* ° TB/1002

J. M. BOCHENSKI: The Methods of Contemporary Thought. *Trans. by Peter Caws* TB/1377

ERNST CASSIRER: Rousseau, Kant and Goethe. *Intro. by Peter Gay* TB/1092

MICHAEL GELVEN: A Commentary on Heidegger's "Being and Time" TB/1464

J. GLENN GRAY: Hegel and Greek Thought TB/1409

W. K. C. GUTHRIE: The Greek Philosophers: *From Thales to Aristotle* ° TB/1008

G. W. F. HEGEL: Phenomenology of Mind. ° || *Introduction by George Lichtheim* TB/1303

MARTIN HEIDEGGER: Discourse on Thinking. *Translated with a Preface by John M. Anderson and E. Hans Freund. Introduction by John M. Anderson* TB/1459

F. H. HEINEMANN: Existentialism and the Modern Predicament TB/28

WERER HEISENBERG: Physics and Philosophy: *The Revolution in Modern Science. Intro. by F. S. C. Northrop* TB/549

EDMUND HUSSERL: Phenomenology and the Crisis of Philosophy. § *Translated with an Introduction by Quentin Lauer* TB/1170

IMMANUEL KANT: Groundwork of the Metaphysic of Morals. *Translated and Analyzed by H. J. Paton* TB/1159

WALTER KAUFMANN, Ed.: Religion From Tolstoy to Camus: *Basic Writings on Religious Truth and Morals* TB/123

QUENTIN LAUER: Phenomenology: *Its Genesis and Prospect. Preface by Aron Gurwitsch* TB/1169

MICHAEL POLANYI: Personal Knowledge: *Towards a Post-Critical Philosophy* TB/1158

WILLARD VAN ORMAN QUINE: Elementary Logic *Revised Edition* TB/577

WILHELM WINDELBAND: A History of Philosophy *Vol. 1: Greek, Roman, Medieval* TB/38

Vol. II: Renaissance, Enlightenment, Modern TB/39
LUDWIG WITTGENSTEIN: The Blue and Brown Books ° TB/1211
LUDWIG WITTGENSTEIN: Notebooks, 1914-1916 TB/1441

Political Science & Government

C. E. BLACK: The Dynamics of Modernization: *A Study in Comparative History* TB/1321
DENIS W. BROGAN: Politics in America. *New Introduction by the Author* TB/1469
ROBERT CONQUEST: Power and Policy in the USSR: *The Study of Soviet Dynastics* ° TB/1307
JOHN B. MORRALL: Political Thought in Medieval Times TB/1076
KARL R. POPPER: The Open Society and Its Enemies *Vol. I: The Spell of Plato* TB/1101 *Vol. II: The High Tide of Prophecy: Hegel, Marx, and the Aftermath* TB/1102
HENRI DE SAINT-SIMON: Social Organization, The Science of Man, and Other Writings. || *Edited and Translated with an Introduction by Felix Markham* TB/1152
CHARLES SCHOTTLAND, Ed.: The Welfare State ** TB/1323
JOSEPH A. SCHUMPETER: Capitalism, Socialism and Democracy TB/3008

Psychology

LUDWIG BINSWANGER: Being-in-the-World: *Selected Papers. || Trans. with Intro. by Jacob Needleman* TB/1365
MIRCEA ELIADE: Cosmos and History: *The Myth of the Eternal Return* § TB/2050
MIRCEA ELIADE: Myth and Reality TB/1369
SIGMUND FREUD: On Creativity and the Unconscious: *Papers on the Psychology of Art, Literature, Love, Religion. § Intro. by Benjamin Nelson* TB/45
J. GLENN GRAY: The Warriors: *Reflections on Men in Battle. Introduction by Hannah Arendt* TB/1294
WILLIAM JAMES: Psychology: *The Briefer Course. Edited with an Intro. by Gordon Allport* TB/1034

Religion: Ancient and Classical, Biblical and Judaic Traditions

MARTIN BUBER: Eclipse of God: *Studies in the Relation Between Religion and Philosophy* TB/12
MARTIN BUBER: Hasidism and Modern Man. *Edited and Translated by Maurice Friedman* TB/839
MARTIN BUBER: The Knowledge of Man. *Edited with an Introduction by Maurice Friedman. Translated by Maurice Friedman and Ronald Gregor Smith* TB/135
MARTIN BUBER: Moses. *The Revelation and the Covenant* TB/837
MARTIN BUBER: The Origin and Meaning of Hasidism. *Edited and Translated by Maurice Friedman* TB/835
MARTIN BUBER: The Prophetic Faith TB/73
MARTIN BUBER: Two Types of Faith: *Interpenetration of Judaism and Christianity* ° TB/75
M. S. ENSLIN: Christian Beginnings TB/5
M. S. ENSLIN: The Literature of the Christian Movement TB/6
HENRI FRANKFORT: Ancient Egyptian Religion: *An Interpretation* TB/77

Religion: Early Christianity Through Reformation

ANSELM OF CANTERBURY: Truth, Freedom, an Evil: *Three Philosophical Dialogues. Edite and Translated by Jasper Hopkins and He bert Richardson* TB/31
EDGAR J. GOODSPEED: A Life of Jesus TB/
ROBERT M. GRANT: Gnosticism and Early Christ anity TB/13

Religion: Oriental Religions

TOR ANDRAE: Mohammed: *The Man and H Faith* § TB/6
EDWARD CONZE: Buddhism: *Its Essence and De velopment.* ° *Foreword by Arthur Waley* TB/5
H. G. CREEL: Confucius and the Chinese Wa TB/6
FRANKLIN EDGERTON, Trans. & Ed.: The Bhaga vad Gita TB/11
SWAMI NIKHILANANDA, Trans. & Ed.: Th Upanishads TB/11
D. T. SUZUKI: On Indian Mahayana Buddhism ° *Ed. with Intro. by Edward Conze.* TB/140

Science and Mathematics

W. E. LE GROS CLARK: The Antecedents Man: *An Introduction to the Evolution the Primates.* ° *Illus.* TB/55
ROBERT E. COKER: Streams, Lakes, Ponds. *Illu* TB/58
ROBERT E. COKER: This Great and Wide Sea: *A Introduction to Oceanography and Marin Biology. Illus.* TB/55
WILLARD VAN ORMAN QUINE: Mathematical Log TB/55

Sociology and Anthropology

REINHARD BENDIX: Work and Authority in In dustry: *Ideologies of Management in t Course of Industrialization* TB/303
KENNETH B. CLARK: Dark Ghetto: *Dilemmas Social Power. Foreword by Gunnar Myrd* TB/131
KENNETH CLARK & JEANNETTE HOPKINS: A Rel vant War Against Poverty: *A Study of Com munity Action Programs and Observable So cial Change* TB/148
LEWIS COSER, Ed.: Political Sociology TB/129
GARY T. MARX: Protest and Prejudice: *A Stud of Belief in the Black Community* TB/143
ROBERT K. MERTON, LEONARD BROOM, LEONARD COTTRELL, JR., Editors: Sociology Today *Problems and Prospects* || Vol. I TB/1173; Vol. II TB/117
GILBERT OSOFSKY, Ed.: The Burden of Race: Documentary History of Negro-White Rel tions in America TB/140
GILBERT OSOFSKY: Harlem: The Making of Ghetto: *Negro New York 1890-1930* TB/138
PHILIP RIEFF: The Triumph of the Therapeutic *Uses of Faith After Freud* TB/136
ARNOLD ROSE: The Negro in America: *The Co densed Version of Gunnar Myrdal's* A American Dilemma. *Second Edition* TB/304
GEORGE ROSEN: Madness in Society: *Chapters the Historical Sociology of Mental Illness. Preface by Benjamin Nelson* TB/133
PITIRIM A. SOROKIN: Contemporary Sociologic Theories: *Through the First Quarter of th Twentieth Century* TB/304
FLORIAN ZNANIECKI: The Social Role of th Man of Knowledge. *Introduction by Lew A. Coser* TB/137